DAZZLE 'EM WITH STYLE

The Art of Oral Scientific Presentation

DAZZLE 'EM WITH STYLE

The Art of Oral Scientific Presentation

Robert R. H. Anholt

W. H. Freeman and Company, New York

Library of Congress Cataloging-in-Publication Data

Anholt, Robert René Henri, 1951–
    Dazzle 'em with style: the art of oral scientific
    presentation/by Robert R. H. Anholt.
        p.cm.
    Includes bibliographical references and index.
    ISBN 0-7167-2583-5
    1. Communication in science. 2. Public speaking.
    3. Lectures and lecturing. I. Title.
    Q223.A63 1994
    501.4—dc20

Printed in the United States of America

1 2 3 4 5 6 7 8 9 0 MB 9 9 8 7 6 5 4

To Sid Simon

Contents

Contents

Foreword

Throughout my many years as a scientist I have been reminded time and again, sometimes painfully, of the importance of the art of oral scientific presentation. The ability to communicate scientific information effectively and powerfully is the cornerstone of every successful scientist's career. In contrast to published manuscripts, oral presentations establish personal contact with an audience. The favorable or unfavorable impression created through this contact often permanently shapes a scientist's reputation—a reputation that can either help or haunt the scientist for the rest of his or her career.

Communicating scientific information effectively is important not only for one's research career but also for teaching others. An effective science teacher must be an organized, articulate speaker who instills in students the spark of enthusiasm that motivates them to learn. Furthermore, scientists as a community have a responsibility to communicate the significance of their findings to the lay public. This is essential if we wish to convince the public at large that ongoing support for science is crucial to maintain and enhance our quality of life. Only those who are effective speakers can explain scientific ideas clearly and understandably.

For all these reasons it is surprising that until now little attention has been paid to instructing young scientists in the art of oral scientific presentation. This splendid book could not have been written at a better time. Robert Anholt is a dynamic lecturer who has trained scores of students to become engaging and confident speakers. In this book he meticulously covers the process of preparing and delivering an oral scientific presentation from start to finish. Dr. Anholt discusses actual presentations from many different disciplines and complements them with pertinent personal examples and anecdotes. Anyone who is currently engaged in giving scientific presentations—or anyone who will be in the future—will benefit from the solid information, helpful advice, and good humor contained in this outstanding book. I am convinced that Dr. Anholt's book will become an indispensable mainstay on every scientist's bookshelf and that it will help all of us to *dazzle 'em with style.*

Peter Raven

Home Secretary of the U.S. National Academy of Sciences and Director of the Missouri Botanical Garden, Peter Raven is an articulate spokesman concerning the world's environmental problems.

Acknowledgments

This book has benefitted greatly from valuable suggestions and perceptive criticisms of many of my colleagues at Duke University. Also quite helpful were those who reviewed my manuscript—Kenneth Curry from the University of Southern Mississippi, Victor Hutchison from the University of Oklahoma, and Carl Thurman from the University of Northern Iowa. I am especially grateful to my dear wife, Dr. Trudy Mackay, for her indispensable help and unabated encouragement. Finally, I am grateful to those undergraduate, graduate, and medical students, as well as post doctoral fellows who read and commented on my manuscript: Nancy Andon, Cristina Davis, Laurel Evers, Marjorie Gurganus, John Kraus, Roberta Lyman, Darin Nelson, Marcia Riggott, and Nina Tang. Certainly they taught me more than I taught them.

Robert R. H. Anholt
February 1994

DAZZLE 'EM WITH STYLE

Introduction

*If you can't baffle 'em with brilliance,
dazzle 'em with style.*

I remember my first scientific presentation. I was a graduate student and had never spoken in public. As vividly as if it were yesterday, I recall the empty room — the place of execution — where I dropped my slides into the carousel with trembling hands. Suddenly, the room was full of people, about thirty of them. I had not even noticed them coming in. There was an ominous murmur and the monotonous sound of the air-conditioning system. It was time. My host introduced me; his two-minute speech seemed to last a century. As he beckoned for me to get up and start my presentation, my heart stopped beating, my feet felt heavy, my hands were sweating. I looked at the dimly lit array of pale, inquisitive faces in the audience, and I managed to extrude the first trembling words in a strange, high-pitched voice that seemed to emanate from a different person. Then the words flowed. I had, however, lost all consciousness and daydreamed about a host of unrelated topics while I delivered my lecture

2

mechanically and automatically. I remember the applause; the next moment the room was empty and I was collecting my slides from the carousel. It was over.

For many of us, giving a scientific presentation can be stressful. Yet the ability to deliver a polished oral presentation in front of an audience of peers is an essential skill that is indispensable for a successful scientific career. Oral communication remains one of the most effective ways by which we exchange information and are introduced to new vistas of knowledge. The skill of presenting an engaging and well-structured seminar often determines our professional reputation and future success—especially when the seminar is part of a job interview. The perception of a thesis defense or a research seminar largely depends on the oratory of the speaker. We are all familiar with the droning genius, offering in a monotonous voice an uninspired monologue directed at the projection screen; potentially brilliant work passes us by as we close our eyes and doze off into the arms of Morpheus. We all recognize the annoying speaker flashing hundreds of slides on the screen one after the other, who is still going on twenty minutes past the allotted time, while the audience slowly sneaks out of the room, leaving behind only the host and the speaker. We all remember those seminars that seemed such a waste of time because "it wasn't even clear what it was about."

Truly memorable presentations occur rarely, but they seem to make up for all the boring, soporific, rambling speeches inflicted on us during our weekly seminar sessions. An engaging, articulate, and entertaining scholar who challenges our interest and projects enthusiasm to the audience opens up a world of intellectual pleasure. The speaker's tantalizing story keeps us spellbound, like children who listen for the first time to the tale of Rumpelstiltskin as he makes the poor miller's daughter guess his name. A talented scientist who at the same time is a skilled lecturer is like Mark Antony telling the people of Rome about the great insights of Aristotle. I am truly grateful to those speakers from whose lectures I have learned so much and benefited so greatly.

Some of us have a gift for lecturing and enjoy public speaking; others dread it. For several years, I have directed the weekly graduate student seminars in the Department of Neurobiology at Duke University. For all first- and second-year graduate students, the seminar is a required course in the art of oral scientific presentation. Every week, one student delivers a presentation in front of his fellow students and the faculty. The student chooses a topic and studies it in great detail through library research and conversations with members of the faculty. The presentation is rehearsed several times in anticipation of the appointed date. During the actual presentation, the student's seminar is evaluated by the audience on forms that rate the quality of the presentation not only in terms of its coherence and logic, but also in terms of delivery and the use of visual aids (Figure 1). These evaluations are discussed with the class at the end of each presentation, and the speaker is expected to improve his or her performance in the next semester by learning from constructive critical comments.

Oral scientific presentation is not only an art. It also is an acquired skill. Few individuals are born brilliant speakers. However, most students can dramatically improve their lecturing skills with experience and proper guidance. During the last several years I have worked closely with scores of students to help them develop lecturing skills, and while few are Mark Antonys, many have become good, if not excellent, lecturers. As with every other skill, general principles underlie the art of oral scientific

Student Seminar Evaluation

Evaluator:

_____ Graduate Student
_____ Postdoc
_____ Faculty
_____ Other

Please answer the following questions.

Background Presentation

Was the problem or question clearly stated? _____
Was the presentation put properly into perspective? _____
Did the student appear to have sufficient knowledge? _____
How could the background presentation have been improved?

_____

Description of the Topic

Was the presentation logically organized and
coherent? _____
Were critical experiments appropriately selected? _____
Were the methods clearly described? _____
Could any of the presented material have been
deleted?_____
Was anything missing from the presentation? Please
elaborate. _____

Did the student appear to have thought critically about the
topic and to have insight into the complexities of the issues
addressed? _____

Were the conclusions clearly stated? _____

Delivery

Did the student make appropriate use of visual aids? _____

Were diagrams, tables, figures, etc., clearly and neatly presented? _____

Did the student handle questions appropriately and comfortably? _____

Could the manner of speaking have been improved?

_____

Overall evaluation:

_____ perfect; no improvement needed of any sort
_____ excellent, but could be improved as noted below
_____ good, but could be improved as noted below
_____ fair; major improvements are necessary, see below
_____ poor; well below average and acceptable standards

Note: A "perfect" rating is only appropriate if you feel that this is the best presentation that you have *ever* heard *anywhere*.

Suggested Improvements:

Figure 1 The evaluation form distributed to members of the audience during the weekly graduate student seminars in the Department of Neurobiology at Duke University. These forms are given to the presenting student following the seminar to provide critical feedback on his or her presentation.

presentation. In this book, I have collected insights and guidelines that have crystallized over years of teaching students and scientists to improve their lecturing skills. I have illustrated my guidelines for oral scientific presentation with real-life examples. I have made every attempt to keep these examples relatively simple without compromising their authenticity, since this book is intended for an audience of young scientists. Should the reader wish to skip over some of these examples, most of the take-home messages may be extracted from the boldface passages of the text. (Although I have attempted to choose examples that will appeal to a diverse scientific audience, the reader may find some unavoidable bias towards my own area of expertise, molecular neurobiology and cell biology.) I hope that this book will appeal to many students and young scientists and that it will help them become confident, engaging, and frequently invited speakers.

1

Preparing a Scientific Presentation

Identify Your Audience

During an election year it is not uncommon for a presidential candidate, dressed in jeans and a safety helmet, to spend the morning addressing construction workers in a Michigan suburb, and the same afternoon, dressed in formal attire, discussing the state of the economy with a group of investment bankers. It would be difficult to imagine the politician wearing the construction helmet during the meeting with the bankers. However unlikely it is that a brick would drop on the candidate's head in the union headquarters during the meeting with the construction workers, the candidate's attire enables the audience to identify and feel comfortable with him or her. Politicians are very conscious of the audience they are scheduled to address and do their homework prior to their scheduled arrangements. Although scientists do not experience the same pressures, it is nonetheless a good idea, when planning a scientific presentation, to investigate beforehand what type of audience is expected. Obviously, it makes a difference whether the audience consists of specialists who all share a common interest with the speaker—as is often the case at specialized symposia organized at annual meetings of scientific societies — or whether the audience shares only a peripheral interest with the speaker — common in presenting departmental seminars at academic institutions, where it is often only the host who is interested in the details of the speaker's work. Does the audience consist of scientifically active Ph.D.s or educated laymen? Does the audience consist primarily of professionals interested in a focused account of accomplished work or of students interested in learning about the speaker's achievements within a wider context? A scientific presentation should always be prepared with the audience in mind. To blindly offer a showcase of your own accomplishments, reflecting only your interests, is a sure recipe for miscommunication and results in a poor performance from the perspective of the audience.

Communication **is the key. Look upon your presentation as a dialogue with the audience, not a monologue.** Be sensitive to the needs and interests of your audience, and reflect on the questions: What do they expect to learn from my presentation? How can my presentation be useful to them? A presentation prepared with these questions in mind is more likely to succeed with the audience than a presentation intended from the outset solely to impress the listeners by glorifying the speaker's self-perceived accomplishments.

In most cases the exact day and time of the seminar are not under the speaker's control, but are determined by the organizer. Given a choice, offer your presentation at the normal day and time set aside for seminars at the host institution; scheduling a presentation outside the routine of the regular audience usually results in lower attendance. When given options about a seminar date at an academic institution, choose to be scheduled neither at the beginning nor at the end of a semester. The audience needs a few weeks to establish its routine at the beginning of each semester; and near the end of the semester, especially when the winter or summer recess is approaching, many people start to suffer from seminar burn-out. The largest and most attentive audiences can be expected in the middle of the semester, when academic life is in full swing. Avoid dates that conflict with student breaks or with major professional conventions. Immediately following the national Neuroscience Meeting, for example, few members of neurobiology or pharmacology departments are interested in listening to yet another seminar. When a regular weekly day for seminars has not been set aside by the host institution, it is wise to choose the middle of the week rather than Monday (when members of the audience are struggling to get back in the mood for work) or Friday (when they are preoccupied with the upcoming weekend).

When given an option regarding the time of day, schedule your presentation at noon, if at all possible. Unfortunately, institutions often schedule seminars at the end of the working day to avoid disrupting other activities. However, many people in their daily cycle experience a natural dip in attentiveness just about that time, and it is not surprising to see people fall asleep at afternoon seminars. Personally, I have to struggle to stay awake at 4:00 P.M. No matter how interesting the topic, I find myself dozing off around 4:20 and waking up again about 15 minutes later. In contrast, most people are at the peak of their alertness at around 11:00 A.M.; noon seminars are therefore preferable, and since lunch constitutes a routine, daily break from work, noon seminars usually do not interfere with other activities at the host institution. (When speaking at lunch time, I always make sure to have a snack about half an hour before so that I will not be distracted by hunger during my presentation.)

13

A key concept in the art of oral scientific presentation is acceptance of the speaker by the audience. The speaker's attitude with respect to the audience often determines whether the presentation will be clouded by an atmosphere of skepticism or received in a welcoming ambiance of motivated interest. Establishing a comfortable contact with the audience should be the first concern of any speaker. Starting off with an anecdote, a good-humored reference to the local football team, or a witty comment that draws smiles from the audience often sets the mood for the remainder of the presentation. Although it may seem a cliché, there is absolutely nothing wrong with the speaker thanking the host for the invitation and the opportunity to present a seminar "in front of such a distinguished audience at this prestigious institution." Flattery works. Most students and professionals identify strongly and proudly with their institution, and a display of respect by a visiting speaker immediately forges a bond with the audience. A gracious expression of appreciation to the host and the institution, followed by a brief anecdote or joke (in good taste), hardly ever fails to break the ice.

Try to find out beforehand who might be in attendance during the presentation. Often it is possible to give credit to a specific member of the audience during the talk. Always greatly appreciated are statements such as "After we learned about the elegant experiments of Dr. Smith [in the audience], we decided to...—" or "Since the approach developed by Dr. Jones worked so well in her system, we adopted a similar strategy," or "Our results agree closely with previous observations by Dr. Doe, who showed—". I remember two incidents in which speakers presented a slide showing a cartoon published in several of my review articles. In one case the speaker, unaware that I was in the audience, did not give any credit at all. In the other case the speaker suddenly realized, while looking at his slide, which credited "Anholt et al.," that I was the Anholt in question. I once heard a speaker present data on calcium influx in synaptic terminals; he was unfamiliar with the most recent publications of one of the pioneers of his field, who happened to be in the audience! Such embarrassing instances can do irreparable damage to an otherwise excellent presentation and are entirely preventable. In these examples, speakers had simply not taken the time to find out who belonged to the departments where they would be speaking and thus were likely to attend their seminars. Invited speakers should always browse through a departmental brochure and try to learn a little about the organization and history of the host institution before arriving on the scene.

Knowing your audience facilitates communication and helps create a comfortable and favorable relationship. In preparing for a scientific presentation, always be concerned with uninformed members of the audience and be aware of these questions: What do I seek to *communicate*? Will they be able to follow me? What will they learn and retain from my presentation?

Structure Your Material

After having obtained as much information as possible about the audience and its interests, the next step in preparing the presentation is outlining the talk. A presentation is usually preceded by a brief introduction by the host and followed by a discussion period. Normally, there is a predetermined time allotted for the entire event. It is crucial for the speaker to stay within the boundaries of this time. **Nothing is more destructive to a presentation than ex-**

ceeding the allocated time. Like hikers who go into the wilderness with a food supply just sufficient for the intended duration of the trip, people who attend a seminar anticipate the predetermined period and come equipped with an amount of listener energy just sufficient to cover this period. As soon as the speaker goes over time, the audience becomes impatient and restless. As a result, the speaker will be forced to rush through the most important part of the presentation, namely the conclusion and take-home message. When the presentation is part of a symposium, the chairperson and subsequent speakers will be greatly irritated if a speaker exceeds the allotted time, since this interferes with the next speaker and delays the entire schedule of the symposium. It is also essential to leave enough time for questions. This provides a precious opportunity for the speaker to engage in a direct discussion with the audience, which is essential when you are trying to communicate information effectively. If a 60-minute presentation is scheduled, prepare a 45- to 50-minute talk. This will leave ample time for the introduction and to answer questions at the end. **A good rule of thumb is to keep the presentation at a length which is 75 to 80 percent of the allotted time.**

The graduate students at Duke University are required to prepare carefully written abstracts of two or three sentences, to be printed below the titles on their seminar announcements. Each abstract must be concise and adequately cover the contents of the talk. The first sentence of the abstract usually introduces the basic question and the perspective of the presentation; the last sentence states the overall conclusion. Sometimes a sentence in the middle briefly summarizes the major experimental findings. **Every speaker, before preparing a scientific presentation, should be able to summarize its content in no more than two or three well-constructed sentences.** This is important for two reasons. First, it insures that you are clearly focused on the major issue and the take-home message of your presentation. If the main issue cannot be clearly explained in a few sentences, a presentation will most likely be diffuse and incoherent. Second, it often happens that one important faculty or group member at a host institution is unable to attend the seminar. Often the speaker has the opportunity to meet with this person individually before or after the lecture, but frequently only for a short time. During this brief period you should be able to explain in a few concise sentences the main issue of your presentation.

I once visited a multinational flavor and fragrance company that had expressed a potential interest in sponsoring olfactory research in my laboratory. The most influential person was the Director of Research, whose busy schedule did not allow him to attend my seminar. I was, therefore, scheduled to meet with him for only 15 minutes later that afternoon. When the time came, he was running behind schedule, and in a few minutes he would have to drive to the airport. My total meeting with him was cut to barely 5 minutes, in which he asked me to summarize my major research findings and the future direction of work in my laboratory. Hardly prepared for this unexpected situation, I gave a rushed and incomplete overview of my research endeavor. Ever since this embarrassing and unsuccessful experience, I have made sure that I can summarize my seminar on the spur of the moment in two or three sentences.

In designing the structure of the presentation, remember the interests and expectations of the audience and put the presentation into context accordingly. Using the same set of data, for example, you can often shift the focus of a presentation from molecular aspects to cellular or behavioral aspects or from the nature and generation of a stimulus to the response of the target tissue. A strong presentation on the effects of bronchodilators delivered via inhalation as aerosols would not be structured identically for a group of physiologists interested in the effects of the drug on pulmonary function and for an audience of biomedical engineers concerned mainly with the design of the inhaler and its mechanism of drug delivery. Similarly, the focus of a presentation on the preservation of tropical rain forests would ideally differ for an audience of zoologists, one of foresters, one of geographers, and of meteorologists.

If it is important to design your presentation according to the interests of the audience, it is essential to make the audience aware of your focus from the very outset. Suspense and mystery are excellent tools for playwrights and movie directors, but they are the nemeses of scientific presentations. To communicate scientific information effectively and maintain the attention of your audience, adhere to the old rule: **"Tell 'em what you're gonna tell 'em, then tell 'em, then tell 'em what you've told 'em."** An outline on the blackboard goes a long way toward keeping the audience aware of the central line of thought of the presentation. A verbal description that outlines the seminar early in the talk also guarantees that the speaker and the audience march to the beat of the same drummer—and in the same direction. "First I will describe to you how this enzyme was discovered. I will then show you evidence demonstrating that this enzyme represents the rate-limiting step in the metabolic pathway under discussion and that it is regulated by calcium. Finally, I will convince you that altered activity of this enzyme as a result of changes in calcium concentration results in abnormalities in bone structure." Such a statement establishes a line of thought that enables the audience to comfortably make an organized mental inventory of the information that is about to follow and to apportion their total listener energy in installments that correspond to the announced segments of the presentation.

Each presentation consists of three segments. (1) The introduction provides the background and perspective necessary to appreciate the remainder of the presentation. (2) The body of the presentation, usually the largest section, conveys new information to the audience; this section can often be divided into distinct, interrelated subsections. (3) The conclusion summarizes the presentation and should provide the audience with a clear take-home message. **When preparing a structure for your presentation, divide the allotted time and assign a defined number of minutes to each section of the presentation—for instance, 10 minutes for the introduction, 30 minutes for the main body of the presentation, and 5 minutes to summarize and conclude.** This provides balance between the different segments of your talk and ensures that the presentation stays within a restricted time frame.

In many cases, the arrangement of slides or transparencies will help structure your presentation by providing landmarks along which the lecture can be organized. In a later chapter we will discuss visual aids in detail; suffice it to mention here that the number of slides should be kept within reason. **A good rule of thumb is to allot approximately 2 minutes of presentation per slide, making 20 to 25 slides a good number to aim for when preparing a 45-minute presentation.** A single sheet of paper with titles or key words that identify each slide and perhaps a few small reminder notes can provide a convenient "cheat-sheet" to which you can refer during your presentation. Alternatively, you can prepare a sheet that contains photocopies of all your slides in the correct sequence. **A well-prepared abstract, an organized set of well-chosen slides, a concise "cheat-sheet," and an outline to put on the board should all help keep you on track during your seminar.**

Know Your Stuff

It happens often, especially in cases of novice speakers, that unjustified self-confidence leads to the belief that eloquence and style will make up for lack of knowledge, incomplete understanding, or absence of crucial data. A false sense of intellectual superiority to the audience, instilled by the assumption that no one else knows more about the topic than the speaker, frequently leads to the illusion that he or she will be able to "wing it" yet give the impression of being on solid ground. More often than not the speaker discovers too late that the audience consists of highly intelligent and insightful individuals. It takes only one knowledgeable listener to expose a lack of knowledge or data that the speaker has tried to hide behind a cloak of superficial information. There is no substitute for knowledge. Long before the question period, it will become evident even to a lay audience whether the speaker has a thorough understanding and a broad, solid command of the field. Nothing is more embarrassing to a professional than to be caught unprepared to discuss recent literature or details of important, albeit peripheral, aspects of the field. Nothing is more disappointing to an expectant audience than a speaker who, having no data or only a limited amount, spends most of the time talking about planned but not yet performed experiments. If you have no data for a scheduled research seminar, choose another topic for which solid data are available or simply decline to speak. **The decision not to speak is sometimes more beneficial to a person's reputation than a lecture devoid of data.**

The extent of a speaker's knowledge reveals itself in subtle ways, especially in the articulation of sentences and the precision of statements. For instance, consider a speaker who asserts the following: "Mammalian pheromones mediate reproductive behavior by interacting with the vomeronasal organ. This chemosensory organ differs from the main olfactory system. It is distinct in all mammals, *except in higher primates, where only a vestigial remnant of this organ is found.*" This speaker displays in a subtle but convincing manner a significant depth of knowledge. The additional information regarding the vestigial remnant in higher primates suggests that the speaker has a broad knowledge of the subject; the term "higher primates" rather than "humans" indicates that the speaker's knowledge is precise. Contrast this example with that of a zoologist who discusses the evolution of birds and mentions that "*Archaeopteryx*, a Jurassic ancestor of birds, was a tree-dwelling creature." Although few members of the audience would challenge this notion, this speaker is perceived as less authoritative than the speaker who states: "Based on the high degree of curvature of the claws of *Archaeopteryx*, which is characteristic of perching birds, it has been argued that *Archaeopteryx* was a tree-dwelling Jurassic bird." In the latter case the speaker backs up the statement that *Archaeopteryx* may have been a tree dweller with morphological evidence, thereby demonstrating knowledge of the literature and of the arguments that support this claim. The careful phrase "it has been argued that" indicates that the speaker is aware of the notion, persistent for many years, that *Archaeopteryx* was adapted to a more terrestrial life style of running rather than perching and dwelling in trees. **Accurate, complete, well-phrased descriptions of scientific information** portray the speaker as a knowledgeable, reliable source of information. In contrast, glib, inaccurate statements that are open to multiple interpretations gradually elicit skepticism and distrust.

Finally, knowledge and data alone are not sufficient. *Critical examination* **of the information is indispensable.** This is perhaps one of the most difficult tasks: to stand back and critically look upon your own work. Yet those who are able to do that can prevent or anticipate embarrassing questions. Furthermore, a critical, careful presentation during which you demonstrate familiarity with pitfalls of experimental design and ongoing controversies in the literature, as well as understanding the limitations of the data presented and their statistical reliability, further instills confidence that you are truly an authority in the field.

Rehearse

No matter how experienced a speaker is, it is always a good idea to rehearse a presentation. Often, the same presentation can be given on a number of occasions, but not without adaptation. "Canned" seminars pose the danger of providing a product not optimally tuned in to the audience — a danger we have already identified. I once listened to a presentation on drug testing given by a representative of a federal drug enforcement agency who evidently no longer rehearsed his lectures. The speaker, in front of an audience of professional chemists, gave the same talk he normally delivered at high schools and community colleges. He proudly presented pictures of a gas chromatograph and an automated analysis system in order to emphasize that these fancy machines really work and give reliable data! To his professional audience, his presentation was disappointing, embarrassing and totally inappropriate. The moral of this anecdote: **Prepare each seminar for every individual occasion *de novo*, always with the specific audience in mind.**

A substantial period of time for preparation of the presentation should be allowed prior to the scheduled date. Frantic last-minute preparations can result in acceptable performances but seldom in memorable ones. A scientific presentation is an expression of creativity, and creation takes time. I usually start thinking about scheduled presentations weeks, sometimes months, in advance. I draw up a rough outline early on and then brood and daydream, letting the presentation go around in my mind, letting the concepts mature at their own slow pace, like a fine wine in the cellar of a French château. Rehearsing the presentation and going through the slides, rearranging them again and again until I am fully satisfied with their final order, is—in my experience—best done in the evening right before retiring. Somehow, the last impressions of the day are firmly embedded and integrated in our minds during our sleeping hours.

A rehearsal in front of an honest and perceptive colleague is invaluable. This person should preferably be someone who could fit in as a member of the prospective audience and who does not feel inhibited about giving frank and critical feedback, with regard to both the presentation's scientific content and the delivery. For such a rehearsal to be useful, honesty must prevail over politeness. What a speaker needs most for the rehearsal is not necessarily a sophisticated expert in the field, but a well-informed colleague with whom a comfortable rapport or good friendship exists. Inexperienced speakers can benefit from rehearsing their presentations with a tape deck or cassette recorder. Listening to your own voice can be very revealing and may help turn a dull, monotonous account into an exciting story. A video recorder can also be helpful and may reveal such distractions as talking to the screen, talking with one hand in a pocket, or talking to the floor.

Prepare — Then Relax

You have familiarized yourself with the composition and interests of your audience. You are armed with solid data, well-rounded knowledge, and a thoroughly organized presentation. You can do nothing more than confidently and quietly await the moment of truth. **Relaxation is now essential.** I have seen more seminars fail because of the self-destructive nervousness of the speaker than for any other cause. Novice speakers tend to experience extreme stress before a seminar. Yet, paradoxically, **to give a good presentation it is essential to be relaxed.** Listening to a nervous wreck is disconcerting to an audience and distracts more than anything from the content of the presentation. Relaxing is, however, easier said than done. I always advise my students to have fun the night before their scheduled presentations, to avoid any further rehearsals or preparations, and to engage in activities that take their minds completely off the upcoming presentation.

The worst period in terms of nervousness is usually the first five minutes of the presentation, during which novice speakers tend to choke, stammer, go blank, or rattle on. After a while, as slides and other visual aids provide landmarks to guide the presentation, the speaker usually relaxes. **To ease into the presentation, a nervous speaker should write down a few opening sentences on a sheet of paper and read them out verbatim in as natural and controlled a voice as possible at the beginning of the presentation (up to the appearance of the first slide), making sure not to speak too fast.** In the end, the best cure for nervousness is experience. Just as time heals all wounds, experience in public speech removes all apprehension. Speakers who have to give important presentations away from home will benefit enormously from practice talks at home in their own departments, lab groups, or journal clubs.

No matter how well prepared a presentation is, unforeseen problems may come up. The projectionist drops your carefully arranged slides on the floor; the light bulb in the projector burns out in the middle of the presentation; a last-minute change of lecture site delays the arrival of the audience. I have waited in front of locked lecture halls while my hosts frantically tried to find the keys. I have spoken into microphones hooked up to defective sound systems. And I have tried to use laser pointers that lacked batteries. These unforeseen setbacks should not rattle you. They are completely out of your control and are the responsibilities of the host and the host institution. Lack of organization here reflects poorly on the host institution, not on the speaker. In these cases, the audience has a responsibility to bear with the speaker and allow extra time to make up for any interruptions.

Yet such events do interrupt the flow of a presentation and are disturbing and distracting. When a laser pointer gives out, you can simply try to ignore the problem by pointing with a pen or a ruler, or describe the items on the slide verbally. Similarly, when a sound system goes down, you can speak louder. In other cases, however — when a crucial slide cannot be shown because of a burnt-out projector bulb — it is often better just to wait until the problem is fixed. Many destructive organizational incidents can be avoided by loading your slides in the carousel well ahead of time. This avoids a last-minute rush that can result in slides put in upside-down or out of sequence. Also remember that some carousels cannot accommodate unusually thick slides, such as those commonly used in Europe. Bringing your own preloaded carousel and your own pointer can prevent many potential problems. Arrive in the lecture room 15 minutes prior to the scheduled time and check that the projector, pointer, room lights, screen, and other accessories are all in good working order. At that point, nothing more can be done to set the stage for a successful presentation, so relax!

Dress for Success

The day of the presentation: here I am in my hotel room enjoying breakfast and coffee brought up by room service. My slides are ready, the presentation is outlined firmly in my mind, and I feel calm, confident, and relaxed. Time to get dressed — but what should I wear? I remember a speaker who interviewed for a faculty position in our department and gave a seminar dressed in a sweaty T-shirt, jeans, and sneakers. This applicant must have thought that in academia how you dress does not matter, only how smart you are. Wrong! **Dressing up for a scientific presentation conveys two important messages: respect for your audience and willingness to conform.** This is especially crucial for job interviews. Many institutions think twice before offering an important position to someone who shows unwillingness to bow to the conventional social graces. Ignoring proper dress codes raises doubts about how the candidate would fit into a team and fears that the candidate's lack of etiquette might embarrass the institution at some future important occasion. It is true that members of an audience may not care about the speaker's attire, but some may be offended. These are often the more senior members, those who make the important decisions! Is it not wise to simply avoid this potentially confrontational issue by dressing conservatively rather than insisting out of principle on making a personal statement of independence through controversial attire?

The small sacrifice of dressing properly shows that the occasion is important to the speaker and demonstrates respect for the host and the host's colleagues. Although most physicians do not need to wear a white coat (unless they are in the habit of frequently spilling their coffee), many patients would feel uncomfortable if their attending physician did not wear one. The white coat is part of the professional image; it instills confidence in the patient about the physician's competence. The same is true for seminar speakers. A well-dressed speaker commands respect as a professional. I always advise my students to dress up for their presentations, and more formal attire invariably correlates with better, more confident, more professional performances.

The formality of attire should be adjusted to suit the group to which you are scheduled to deliver your presentation. For a man, when speaking at a conservative grant-awarding foundation in Chicago, it might be wise to dress in a three-piece suit, whereas a sports jacket and a colorful tie may fit in better when addressing a small, less conservative biotechnology company in California. Jacket and tie may be unnecessary at an informal Gordon Conference in New Hampshire, but a clean, pressed shirt is perfectly appropriate. Black tie would not be out of place for a keynote address following a banquet.

Women have a wider selection of dress options than men. Just as for male speakers, however, conservative attire is always appropriate and ensures that attention remains focused on the presentation rather than on the speaker. Women should also be aware that at some occasions a radio-microphone may be used. These obnoxious instruments consist of a clumsy little box that is supposed to fit in a pocket and a microphone with a clip designed to be fixed onto a lapel or tie. I have seen many female speakers, whose dresses did not provide places to accommodate these items, suffer the inconvenience of having to hold the box in one hand and the microphone in the other throughout their presentations.

Finally, the moment is here! You are entering the lecture hall, well-prepared and appropriately dressed, and you take a moment to survey the arena. "Let the games begin."

The Structure
of a Scientific
Presentation

The Title:
Information in a Nutshell

The first concern when preparing a scientific presentation is its title, which provides the first indication of the presentation's perspective. It should be brief and accurately cover the content of the presentation. Whereas *Gone with the Wind* is an attractive title for a novel or movie, "The American Civil War as Experienced by an Aristocratic Southern Woman" would be more suitable for a scientific presentation. Some scientists, like advertising agents, attempt to inject some humor into the titles of their presentations. "Thoughts of a Calcified Biologist" was the title of a seminar on aspects of calcium metabolism; however, humor lasts only a short while, and there is no information in the title to put the presentation in perspective.

Your title should help set a context for your presentation. The title should be concise and general enough to appeal to a wide audience, but not so general that it loses meaning and overstates the actual content to such an extent that the seminar itself will be anticlimactic. Thus, titles such as "Consciousness and the Mind" or "Expansion of the Universe" are best avoided. Often a general title can be qualified by a subtitle that specifies the scope intended for the actual presentation. "Axonal Regeneration in the Central Nervous System: The Role of Growth Cone Proteins" or "Global Warming: Its Effect on Speciation of North-American Songbirds" are titles that describe topics of general interest in the main heading and narrow the scope of the presentation (within this general topic) in the subheading. Note that both the main heading and the subheading are brief, to the point, and catchy. Verbs, which turn the title into a statement, should be avoided; "Growth Cone Proteins Play a Role in Axonal Regeneration in the Central Nervous System" is too long and has lost its brisk and striking character. The old scholarly preposition "on" — "On the Role of Growth Cone Proteins in Axonal Regeneration" — now seems pedantic even for written manuscripts and is certainly too formal for oral presentations.

Context and Perspective: Zooming In

A bird's eye view of London: the Tower, Westminster Abbey, Big Ben, and the Houses of

Parliament on the Thames. The camera zooms in near the river, where a group of people listen to a political speech. Suddenly the body of a woman strangled with a necktie floats onto the shore. In the next scene, the camera shows a bird's-eye view of Covent Garden. We zoom in further to the vegetable stalls in the market, where the camera focuses on one man leaning against a fruit stand and eating an apple. It is clear that he is one of the main players, the criminal, in the movie. These scenes from *Frenzy*, one of Alfred Hitchcock's best, set the stage during the opening credits of the movie. We know the location, the approximate era, the crime, and the main player, so the story can take off. Another cinematographic masterpiece is Federico Fellini's *La Strada*. Its most impressive scene is the last of the movie, where the strong man Zampano, a traveling circus performer, finds himself a human outcast after he has murdered his rival and caused his devoted female companion to wither away from grief. He walks into the ocean, looks toward the horizon, and falls onto his knees in tears. The camera slowly zooms out as the music swells, showing the vast beach and ocean and the ever-smaller figure of Zampano: a small man in a vast universe.

Zooming in and zooming out are not only effective cinematographic tools. They are essential concepts in structuring a scientific presentation. In fact, **zooming in is the *only* effective method to put a presentation into perspective. The presentation must always start with the description of an important general principle, then gradually focus in from there onto the experimental model that the speaker wishes to describe.** Talking about molecular aspects of olfaction, for instance (one of my own great interests), I always start my presentation by mentioning that "the olfactory system mediates the molecular recognition of a vast variety of molecules." I further mention that "these molecules represent chemical signals from the environment that provide information about the localization of food and the availability of reproductive partners and are, thus, essential for the survival of most animals." These opening lines make the importance of the topic unambiguously clear in its most general sense. It is now easy to pose the major overall question on which the rest of the topic will focus: How does the olfactory system mediate the discrimination of myriads of odor molecules? The next step is to explain that it is essential to study this problem at the molecular level—that is, at the level of odorant receptors and the transduction processes that they trigger. But I do not immediately jump into this topic without explaining to the audience the functional anatomy of the system. I zoom in gradually, step by step. First I discuss the histology of the olfactory system and explain the organization of the tissue. Then I descend to the cellular level and explain the structure of the olfactory neuron. Finally I arrive at the molecular level and provide an account of what is known about the molecular components of the chemosensory membrane: the facts that underpin the work I am about to discuss. By zooming in I provide the audience with a frame of reference, a bird's-eye view. I arouse their interest by starting with a general principle and narrowing the focus to the point at which they can understand and appreciate the experiments about to be described.

Another example of contextualizing a presentation by zooming in might be a paleontologist analyzing fossil specimens of Miocene apes. It is of paramount importance for the speaker to convince the audience from the beginning of the presentation that the analysis of these fossils is not merely incidental but in fact is crucial to our understanding of primate evolution. To achieve this objective the speaker would first provide an overview of primate and especially human evolution and indicate that there are significant gaps in our understanding of the evolutionary interrelationships between *Homo sapiens* and the African apes. The speaker would follow with a description of the similarities and differences in cranial features and blood serum proteins between orangutans on the one hand and chimpanzees, gorillas, and humans on the other, and present the argument that the African apes must share a common ancestor with modern man. This sets the stage for an account of the discovery of *Sivapithecus* and *Ramapithecus* found in the Chinji and Nagri formations in Pakistan, respectively, and proposed to represent the first Miocene fossils closely related to a common ancestor of apes and hominids. The speaker can now continue to discuss in detail how the fossil bones of *Sivapithecus* and *Ramapithecus* reveal traits preserved to greater or lesser extent in each of the divergent descendants. A comparison can then be drawn between the Pakistani specimens and similar fossils found more recently, in 1989 in Greece (*Ouranopithecus macedoniensis*) and in 1992 in Hungary (*Rudapithecus hungaricus*). The description of these European specimens prepares the audience for a discussion of whether humans are more closely related to gorillas or to chimpanzees. The speaker can now zoom out again, summarize the different possible evolutionary trees, and conclude that the discovery of the Miocene fossils in Pakistan and Europe have brought us closer to the identification of the common ancestor of ape and man. If the detailed comparative morphology of the various skulls and bones were described for its own sake without having been placed in the context of human evolution by zooming in and zooming out, the presentation might bore the audience to tears. In contrast, to submit these findings in the context of a major biological question—namely, the quest for the evolutionary origin of man—makes the presentation fascinating.

I remember one speaker who discussed the mechanism by which the flow of calcium into adrenal chromaffin cells triggers the secretion of adrenaline. The speaker explained that calcium enters the cells via two different pathways: the usual calcium channel, which is opened as a result of changes in membrane potential, and a second "facilitating" calcium channel. The detailed pharmacological description that followed showed how these calcium channels were regulated. Although the work was scientifically superb, the audience was disoriented from the very beginning of the seminar, since the speaker failed to provide an outline that indicated why the work was important and what question the seminar was meant to address. Only at the end of the presentation was it evident that the facilitating calcium channel enables rapid and massive degranulation of the cells, which results in a fast increase in adrenaline levels in the circulation. Although it was not clearly stated, the speaker had in fact discussed an important molecular control point of the "fight or flight" response, a behavior essential for survival. If he had started off with a description of the role of adrenaline in mediating the "fight or flight" response, thus informing his listeners that regulation of the facilitating calcium current represents the molecular pathway that enables this response, then the audience would have appreciated the importance of the detailed characterization of this channel. The speaker made a crucial mistake: he zoomed out without ever having zoomed in! Although ultimately he did provide a punchline, failure to place his presentation in perspective from the very beginning caused him to lose his audience's attention, and this seminar did not do justice to his otherwise outstanding work.

Zooming in has two important advantages. First, it emphasizes to the audience that the work to be described bears relevance to an important scientific principle rather than being an insignificant, isolated contribution. Second, zooming in defines the intellectual borders of the presentation. The scope of the talk is limited by the diameter of the diaphragm of our zoom lens. We cannot easily stray outside this territory. Thus, zooming in automatically helps us focus the presentation and support its coherence.

The process of zooming in can follow a historical description. For example, neurons depend on growth factors for their proliferation and survival. When discussing neurotrophic factors, a historical account of the exciting and Nobel Prize-winning discovery of nerve growth factor by Rita Levi-Montalcini in the 1960s might be presented. This can be followed by the discovery in 1990 of other related neurotrophic factors, such as brain-derived neurotrophic factor and neurotrophin-3, which affect the growth and differentiation of different classes of neurons. Following the historical development of this field demonstrates how the excitement of the discovery of nerve growth factor as the first neurotrophic factor received additional impetus from the subsequent realization of the existence of a whole family of related factors, of which many members may still await discovery.

Similarly, under certain conditions genes can move around in the genome, a process requiring pieces of DNA known as transposable elements, which can be exploited to generate mutations. A presentation on mutagenesis through transposable elements could begin with a description of the classic and elegant experiments by Barbara McClintock that led to her discovery of "jumping genes" in maize. This revolutionary concept, which antedated modern molecular genetics, was underappreciated for many years; finally she received appropriate recognition in the form of the Nobel Prize. Such historical accounts are also processes of zooming in and allow the audience to evaluate the importance of the speaker's contributions relative to those of predecessors and within the context of the major scientific principle.

There are cases in which the historical perspective becomes an absolute necessity for the context of the presentation. Consider a lecture that deals with predictions of seismic activity along the San Andreas fault. Great earthquakes (those whose magnitude exceeds 8.0 on the Richter scale) tend to occur in irregular cycles.

For instance, the San Francisco Bay area experienced several small earthquakes in 1836, 1838, 1865, and 1868 preceding the great earthquake of 1906. After a period of quiescence, the area came to life again in the 1950s. Small earthquakes along the fault in 1974, 1984, and 1988 preceded the major Loma Prieta earthquake that caused the collapse of the Bay Bridge on October 17, 1989. To put a presentation on predictions of seismic activity along the San Andreas fault into perspective, a speaker must not only explain the tectonic movements that generated the fault and its geographical extent, but also give an account of the dates and locations of previous earthquakes. Without this historical perspective, the audience cannot appreciate the complexity and significance of the speaker's predictions.

When putting a presentation in historical perspective, always give appropriate credit to contributions by others in the field. This is important for two reasons. First, one of the major contributors (or a friend familiar with the work) may be a member of the audience and would greatly appreciate being cited. Second, the audience develops sympathy for a speaker who generously gives credit to previous investigators, colleagues, and coworkers; conversely, it distrusts a speaker who never mentions anyone else by name, for the sake of appearing to be the sole player in the game. I remember a speaker who interviewed for a faculty position at a major university. His presentation was spectacular, state-of-the-art molecular neurobiology. But he did not refer at all to other groups who had made fundamental contributions to the same field several years earlier. Nor did he give explicit credit to his colleagues, who contributed to his achievements in a major way. The main publication describing his work listed him as third among five authors, and he did not make clear which contributions were his and what percentage of the work was done by his coworkers. His urge to make a big impression and his mistaken idea that withholding credit from his fellow researchers would help him succeed had an adverse effect, generating a feeling of distrust despite the excellence of the work. If he had been more generous during the introductory segment of his presentation, his presentation would have been much better received and might have led to the desired faculty position.

Telling a Story

"When I go to a seminar, I want to hear a story." This simple remark by a friend and colleague of mine captures the essence of a successful scientific presentation. **There is a distinct difference between summarizing a collection of facts and telling an exciting and interesting story.** A story has a perspective, a context, a plot, and a climactic conclusion. A story should keep the listeners spellbound and fascinate them while the plot unfolds.

To deliver a coherent presentation you must have a focused concept of the topic in mind. **The ability to speak coherently is closely correlated with the ability to *think* coherently. A clear thinker separates the central, relevant issues from merely supportive peripheral information and will not allow the direct line of thought to be interrupted by sidetracks.** Moreover, experimental results do not necessarily emerge chronologically according to their scientific significance. When presenting data, it is, therefore, essential to arrange the observations in a sequence that makes sense to the uninitiated members of the audience in order to generate a coherent story. I remember a semi-

nar that dealt exclusively with glutamate receptors, yet completely lacked focus. It consisted of
a collection of pharmacological, biochemical,
and physiological facts that were not clearly in-
terconnected, and the vast amounts of irrelevant detail lulled the disoriented audience to
sleep. The speaker had not invested the effort to
think about what was the central question to be
addressed. Since glutamate receptors happened
to be a fashionable topic, it was taken for
granted that the audience *a priori* would be interested and assumed that an indiscriminant
listing of facts would automatically result in a
satisfactory presentation.

**A story should have one focus and convey
a single major message.** Never break up a
seminar by addressing one topic and then
"switching gears" to address a different and unrelated topic. This often happens when a
speaker does not have sufficient data to fill the
allotted period of the seminar and tries to divide
the available time into two portions, each dealing with different projects, both of which are
often inconclusive. Sometimes young scientists
interviewing for a faculty position think that
combining more than one topic in a single seminar will show off the breadth of their scientific
exposure and enable them to display as many of

their past accomplishments as possible. Readjustment to a new topic, however, requires considerable listener energy. After a listener has invested a considerable amount of attention in following one story line, it is disconcerting to be returned to "square one." Expecting to hear only one seminar, the listener feels tricked into yet another. Changing the subject midway through a seminar makes the audience feel as if it is watching a Shakespearean play that suddenly continues as Greek tragedy, or as if they are attending a rock-and-roll concert that unexpectedly features Schubert lieder. The speaker will give a much better presentation by choosing to speak about only one of two beloved topics, even if it means shortening the presentation (a tactic often appreciated by the audience). Sometimes it is better to describe an exciting project that was completed rather than address incomplete current work; at another time a speaker may prefer to discuss exciting research in progress and forget about work done in the past that has been thoroughly chronicled and lost its appeal of innovation. Whatever the case may be, a scientific presentation should address a single topic and explore one focal idea.

Once a pediatrician gave a seminar to a group of basic scientists on inherited diseases that arise from defects in lipid metabolism. Determined to cover all the different diseases of this kind that he had ever encountered—GM1-gangliosidosis, Tay-Sachs disease, Hurler's disease, Gaucher's disease, and Niemann-Pick disease—he provided very little introductory perspective and simply devoted about 10 minutes to the description of each syndrome. The presentation was an interesting catalogue of disorders, but it was not a *story*. It would have been a far more interesting seminar if, at the beginning, the speaker had stressed that a defect in a single enzyme, perhaps arising from a single base substitution at the level of the DNA, could make the difference between health and tragic illness. He could have mentioned that this is exemplified most dramatically by lysosomal storage diseases, especially those that affect lipid metabolism. He could then have provided some essential background by describing the biosynthesis and degradation of gangliosides and cerebrosides. Subsequently, he might have focused on only one of the best documented diseases, such as Tay-Sachs disease, an autosomal recessive disease that has an especially high incidence among Jews of Eastern European descent. Tay-Sachs disease arises from deficiency of a specific β-N-acetyl-hexosaminidase, resulting in accumulation of ganglioside GM2, normally present in the brain in only trace amounts. The speaker could have used Tay-Sachs disease to illustrate how a single deficient enzyme can have devastating consequences, including dementia, motor disorders, blindness, and usually death before the age of three. He might have described diagnosis and management of Tay-Sachs disease and finished his seminar by discussing the feasibility of a future cure for this and related disorders through gene therapy. If he had provided a proper perspective and focused on one major disorder in great detail, from its diagnosis to possible future treatments, his seminar would have elicited far greater interest and appreciation.

One of the best presentations I ever heard was given by one of my students, whom I'll call Joe. Joe's presentation described the discovery that the high-affinity receptor for nerve growth factor and the *trk* proto-oncogene product, which phosphorylates target proteins on tyrosine residues, are the same molecular entity. This suggests that activation of nerve cells by nerve growth factor is triggered via tyrosine phosphorylation. Joe first described the characterization of nerve growth factor and its receptor. He then "switched gears" and talked about the *trk* proto-oncogene. Finally, he brought the two topics together and showed that both groups, the neurobiologists working on the receptor for nerve growth factor and the oncologists working on the *trk* proto-oncogene, realized that they were studying the same protein—a major discovery! Joe's seminar was a tour de force. He told a fascinating story that unfolded to an exciting climax. Superficially he appeared to be talking about two separate subjects (the characterization of the receptor for nerve growth factor on the one hand and studies on the *trk* proto-oncogene on the other), but the seminar was really based on one focal idea and it conveyed a single important message.

Another delightful scientific story was presented by another one of my students, Nina, who discussed the mechanism of programmed cell death, a phenomenon known as apoptosis. First Nina convinced the audience that the question of programmed cell death was a universally important biological issue. She explained that apoptosis is an organized cellular process distinct from the traumatic cell death that occurs during necrosis. She then zoomed in to the nematode *Caenorhabditis elegans* and explained that this simple worm offers a beautiful model system because of its small number of individually identifiable cells and potential for genetic manipulation. She went on to describe a number of mutants that were defective in programmed cell death and identified one particular gene, *ced-9*, as a gene that normally suppresses apoptosis. Like Joe, she then apparently "switched gears" and focused the audience's attention on programmed cell death in a variety of mammalian systems. She showed that a protein originally identified in a human B-cell lymphoma, known as the product of the *bcl-2* proto-oncogene, suppressed apoptosis in mammalian cells in a way similar to the product of the *ced-9* gene in *C. elegans*. At the end of her presentation, she described an experiment in which DNA that encodes the mammalian *bcl-2* protein was introduced into *C. elegans* and was able to suppress programmed cell death in the nematode! Thus the descriptions of *C. elegans* and mammalian systems came elegantly together as her seminar "zoomed out" to an exciting and climactic conclusion of general biological importance—namely, that nematodes and mammals are likely to share a common universal pathway for programmed cell death. Like Joe's presentation, Nina's seminar presented a *story*, not just a presentation of experimental facts.

To construct the plot for a scientific "story," it is often helpful to phrase the basic idea underlying the talk as a question. The seminar then becomes structured as the gradual unfolding of the answer to that question. Writing the question on the board or projecting it on the screen helps provide an unambiguous focal issue placing both the audience and the speaker on the same wavelength. Basing your seminar on a single, well-formulated question results in a presentation that almost by necessity focuses on one basic issue. Asking fundamental questions is the essence of science. Therefore, virtually every scientific presentation can be structured as an answer to a fundamental question. These questions can be broadly or narrowly focused. How do neurons from the lateral geniculate nucleus project to the visual cortex? How does the extracellular matrix influence the growth of axons? How does deforestation cause climate changes in the Northern Hemisphere? How is fatty acid metabolism regulated during hibernation? The structure of the question can also help place a presentation in proper perspective for a particular audience. Consider the questions How will research on AIDS benefit future AIDS patients? and Will future research on AIDS benefit the American economy? These two questions would support two distinct seminars dealing with the same issue, namely, benefits of AIDS research. I often advise my students to make it clear at the beginning of their presentations what question is being addressed.

State the major question of the presentation at the beginning and then break this question down into subquestions arranged hierarchically to gradually unfold the answer. Throughout this process, you can ask a question, provide an answer to that question, and then formulate the next question based on the previous answer. Segmentation of your talk into a sequence of chapters that are logically related to one another and that step-by-step disclose the answer to the underlying question of the presentation is a powerful device for structuring a coherent lecture.

In 1982, for example, I attended a student's thesis defense that focused on the structure and function of the nicotinic acetylcholine receptor, the first neurotransmitter receptor to be purified. After acetylcholine is released from a nerve ending at the neuromuscular junction, it interacts with this receptor on the postsynaptic membrane. This interaction generates a flow of ions across the muscle membrane, which ultimately leads to contraction of the muscle. In the late 1970s a controversy existed as to whether the purified acetylcholine receptor also contained the acetylcholine-activated ion channel. The student began his thesis presentation by making the audience aware that the acetylcholine receptor represents the key component that mediates synaptic transmission at the neuromuscular junction and that understanding structural and functional relations within the receptor molecule is therefore essential to understanding the molecular basis of neurotransmission. This convinced the audience that the topic was important and placed it in perspective. The general question underlying the student's seminar was: How is it possible to study structure and function relationships in the purified acetylcholine receptor molecule? The overall answer to this question was: It is possible by incorporating purified receptors in artificial membranes under conditions in which these receptors remain intact and retain acetylcholine-activated ion channel activity.

During the seminar the answer to the overall question unfolded in a stepwise fashion, a process during which the question was broken down into a sequential series of smaller questions. To ensure that the audience was able to follow the logic of the presentation, each of the questions appeared on a slide and was addressed with data from appropriate experiments; afterward, another slide formulated the conclusion. The first question was: Can acetylcholine receptors be purified under conditions in which ion channel activity remains preserved after reincorporation into artificial membranes? Experimental data showing that this indeed is the case were summarized on a slide with a conclusion written on it. The next question appeared on another slide: Are the binding sites for acetylcholine and its associated ion channel part of the same protein? Evidence was presented showing that receptors purified under conditions defined in the first segment of the presentation can mediate the flux of ions when activated by acetylcholine. A summary slide stating "The purified acetylcholine receptor contains both the acetylcholine binding sites and the acetylcholine regulated ion channel" concluded the second segment of the presentation. This was followed by the final subquestion: Are the pharmacological properties of the acetylcholine receptor in artificial membranes similar to those in the native membrane? Data were presented demonstrating that after incorporation in artificial membranes, acetylcholine receptors are pharmacologically intact and show the same acetylcholine-induced conformational transitions as receptors in the native membrane. This third segment was followed by an overall summary slide which stated that pharmacologically intact acetylcholine receptors could be purified and reconstituted in model membranes, and that this approach demonstrated that the purified acetylcholine receptor molecule contained both the binding sites for acetylcholine and the acetylcholine-regulated ion channel. The clear formulation of the major question (and its answer) as well as its breakdown into an organized series of sequential secondary questions and answers provided a logical structure for the presentation and made it easy for the audience to keep track.

Another example of a coherent presentation that was structured as a hierarchically arranged series of questions was a seminar given by a student in ecology. The student discussed mathematical models that could describe the growth of populations. The basic question addressed by the seminar was: Can the growth of a simple population that consists of a single species and lives in a controlled environment be predicted mathematically? Since the answer to this question appeared not to be a simple yes or no, the student proceeded to break this question down into a series of subquestions. The first was: What mathematical equation might describe the growth of such a population? The student explained that the size of the population would be measured after each generation and that, rather than counting the actual number of individuals, the population size would express the percentage of some limiting maximum number of individuals. Thus, the percentage of the population after n generations, P_n, could be described by the simple equation $P_{n+1} = kP_n (1 - P_n)$, where $0 \leq P_n \leq 1$ and where k is an environmental constant that depends on ecological conditions such as food supply or space limitations. The student wanted to argue that this seemingly simple equation could generate complex behavior depending on the value of k. Thus the next question was: How does this system behave at low values of k? The student answered this question by showing an iterative model of the population's size at a value $k = 0.5$ and demonstrated that, at this value of k, the population would die out after about 8 generations.

To make the model more interesting, the student now proceeded to ask the contrasting question: What happens to the growth of the population at larger values of k? A series of intriguing observations was now presented, showing that when values of 1.2, 2, or 2.7 were assigned to k, the size of the population would stabilize at a defined limiting value. However, when the value of k was set at 3.1, the size of the population showed fluctuations between two distinct values, and at $k = 3.7$ a more complex pattern of oscillations in population size between four distinct values was observed. Finally, the student asked whether conditions might exist under which the behavior of the population became unpredictable. The answer to this question was revealed when he entered a value of 4 for k in the equation. The model now showed random fluctuations in the size of the population, revealing inherently unpredictable behavior. After this critical analysis of the model and its implications, the student concluded that the dynamic behavior of a simple population under controlled environmental conditions can be predicted only under certain conditions that depend on the value of the environmental constant, k, and that the simplest prediction, namely extinction of the population, can be made reliably only when k is small. Focusing in on a single model to address an important biological issue by examining the model via a hierarchically arranged series of

questions allowed the student to provide a coherent analysis that was intelligible even to a
nonmathematically inclined audience.

Finally, let us consider an example from the
industrial world. A university professor has developed a novel transdermal drug delivery system that is particularly suitable for the sustained administration of steroids. Realizing
potential financial benefits, she attempts to convince a large pharmaceutical firm specializing
in steroids to market her invention. The success
and financial future of the college professor
hinges on a presentation she has to give to the
board of directors of the pharmaceutical company. Her goal is to convince them that use of
the new drug delivery system would increase
their profits. For the presentation to be convincing, it has to be focused and coherent. Once
again, it is possible to set up the talk as a series
of questions. The major question would be:
"How can the large pharmaceutical company
benefit from the drug delivery system developed by the college professor?" The professor
will explain that the system developed in her
laboratory can be used to deliver pharmaceuti-
cals produced by the firm more effectively,
thereby providing an advantage in the market.
The overall question can be subdivided into

smaller questions: "What type of pharmaceutical formulations could be developed?" The professor will indicate how certain products could be developed only as a result of combining her expertise with the resources of the large company. "How much profit could be expected from marketing the proposed products?" would be the next question. She will now point out the size of the potential market, the ease with which already licensed compounds encapsulated in a new non-invasive delivery system would be approved by the Food and Drug Administration, the scarcity of competitors working on similar products, and so forth. At the end of this analysis, a reasonable estimate of profits envisioned to be received by the large pharmaceutical firm and royalties payable to the professor would be discussed. Finally the important concluding question is asked: "How much would the pharmaceutical firm have to invest?" The professor would indicate to the board of directors that the necessary investment would be modest for a company of their size. The presentation can now be summarized by concluding that it would be advantageous for the large company to market the researcher's invention. Structuring the talk as a hierarchical series of questions has resulted in a focused and convincing presentation.

Mainstream and Sidetracks

The most characteristic feature of an exciting story is its *momentum*. Momentum is hard to define. It is the perceived integrated quality of the presentation, composed of the inherent interest of the story, its structure, and its delivery. A story that lacks momentum puts an audience to sleep, whereas a story endowed with it keeps the listeners at the edge of their seats. An unfocused story, not properly put into perspective, can never gain momentum, but focus and careful structure of the presentation alone are not sufficient. The delivery, which we will discuss in greater detail in a later chapter, contributes in a major way to the momentum of a presentation. However, focus, coherence, and a well-orchestrated plot are prerequisites for the seminar to acquire momentum, which may ultimately generate that conquering pizzazz and irresistible flair that make a presentation truly memorable.

As I have asserted, a scientific presentation must have a focal point. The mainstream of the presentation should be directed toward answering the single major underlying question. However, it is often necessary while building an argument to provide the audience with background information that is not directly relevant to the major focus of the presentation, but nevertheless essential for the audience to appreciate the argument thoroughly. Several such excursions may be required during the course of a normal presentation. These sidetracks have their advantages and inherent dangers. On the one hand, they provide accentuation and depth to a presentation in the same way in which the multitude of colors of Renoir's marvelous painting *La Fête dans le Moulin de la Galette* provides an unequaled feast of light, or the embellishments of a Mozart aria provide richness and glamour to the music. But more often than not, sidetracks to the mainstream of the presentation threaten the loss of the lecture's most important attribute — its momentum.

There are three simple rules for preventing the loss of momentum as a result of sidetracks: (1) Keep the number of sidetracks to a minimum and use only those that are absolutely essential. (2) Keep the excursion from the mainstream as brief as possible, providing only the minimal amount of ancillary information that is absolutely crucial for a full appreciation of the presentation's mainstream. (3) Always make clear where the sidetrack

starts and, when it is complete, return to the same point of the mainstream (Figure 2).

When giving a seminar on the neurobiology of learning, for example, one might build the seminar around the argument that the phenomenon of "long-term potentiation" in the hippocampus underlies at least some forms of learning. After describing the phenomenon of long-term potentiation, it may be essential to familiarize the audience somewhat with the pharmacology of glutamate receptors, which mediate long-term potentiation by allowing calcium to flow into the postsynaptic cell. The pharmacology of glutamate receptors, although not part of the mainstream of the presentation, may constitute a sidetrack that provides ancillary information necessary for the remainder of the presentation. Since it would be very easy to give a whole seminar on the complex pharmacology of the glutamate receptor alone, the speaker must limit the description to the minimal information needed to support the rest of the presentation. To introduce this sidetrack, the speaker, while showing a slide demonstrating the phenomenon of long-term potentiation, could say, "At this point I have to take a few moments to describe to you the molecular basis for this phenomenon, namely the glutamate

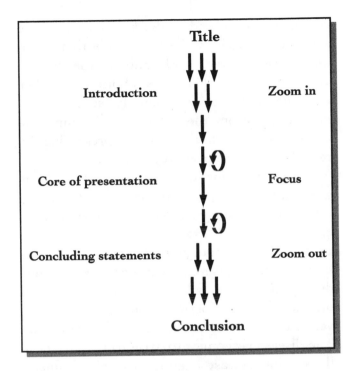

Figure 2 Diagrammatic representation of the flow of a focused scientific presentation. The circular arrows indicate sidetracks, which should always be short and return to the same point in the mainstream.

receptor." At the end of the *brief* sidetrack, the speaker could show the same slide that introduced the sidetrack and continue the presentation from there. This allows both the speaker and the audience to mentally separate the mainstream from the sidetrack, retaining the emphasis on the mainstream and preserving the momentum of the presentation.

The temptation to enter into sidetracks is greatest with interdisciplinary subjects or topics that bear directly on diverse areas of biological sciences. For example, let us consider a presentation on malignant hyperthermia. This human genetic disorder only becomes manifest during anesthesia, when it results in skeletal muscle rigidity, hypermetabolism, and high fever. If not immediately reversed, it can be fatal. The disease is thought to arise from a defect in the calcium-release channel of skeletal muscle sarcoplasmic reticulum, also known as the ryanodine receptor, which is encoded by the *RYR*1 gene. This gene has been localized to human chromosome 19q13.1. A presentation on malignant hyperthermia for a group of clinicians could focus on the function of the normal gene product and explain how a genetic defect in the *RYR*1 gene results in impaired calcium metabolism in skeletal muscle during anesthesia. The speaker may want to emphasize that knowledge of the chromosomal localization of the *RYR*1 gene may enable the development of improved diagnostic tests to identify individuals predisposed to malignant hyperthermia. It may be necessary for the speaker to enter into a sidetrack to describe the genetic mapping procedures. But by showing large numbers of families containing normal and affected siblings and by detailing all the chromosomal markers and linkage studies, a speaker could easily allow this type of sidetrack to overtake the focus of the presentation. For this audience the brief digression could start at the point where the lecturer has indicated the problems which may arise during anesthesia in patients afflicted with malignant hyperthermia. At the end of the sidetrack the speaker can return to the same point and indicate how information on chromosomal localization of the *RYR*1 gene may be of value in identifying the disorder in individuals at risk prior to administering anesthesia.

The same speaker addressing an audience of geneticists involved in mapping the human genome, however, might well want to focus the presentation primarily on the mapping procedures and describe in great detail several linkage studies and polymorphisms in and near the *RYR*1 gene. The chromosomal localization itself rather than the manifestation of the disease can now be placed in the spotlight; in this configuration, the description of symptoms resulting from the genetic aberration becomes the sidetrack, serving to illustrate how mutant phenotypes encountered in certain families can be identified and used to map the gene. The description of the disease and its prevention, which represented the main focus of the presentation for the audience of clinicians, becomes merely incidental to the audience of geneticists who are interested in the organization of the long arm of chromosome 19. Notice how in this example the composition of the audience determines the focus of the presentation and how this focus can shift dramatically even though the central topic remains the same.

It may happen that during the course of a presentation the speaker realizes that he or she forgot to mention something that should have been said earlier. This can pose a problem, since backtracking is one of the most certain ways to undermine momentum. The best way to avoid backtracking is to be so well prepared for your presentation that the problem will never arise. If it does occur, however, you must carefully and quickly decide how important the deleted material really is. Is it possible to simply carry on without it? This is often the preferable solution. For instance, let us consider a speaker who discusses the function of the *Hox* gene, which regulates segmentation during ontogenesis of the mouse. This speaker intended, but forgot, to mention that this gene was discovered as a homologue to the *Ubx* gene, which controls segmentation during the development of *Drosophila*. If this information is not essential for the rest of the presentation, which focuses on embryogenesis of the mouse, it can simply be deleted. The predicament is slightly more complicated if several *Drosophila* researchers working on the *Ubx* gene are members of the audience. The speaker must now decide whether to offend them by not giving credit to *Drosophila* researchers, who opened up the field currently being discussed, or whether to harm the momentum of the presentation by backtracking (notice also the temptation for sidetracks, which in this case could involve a lengthy discussion of segmentation in *Drosophila*). If the deleted information is crucial for the final conclusion of the talk or for other reasons, the speaker has two options: (1) to incorporate the information somehow at another place in the presentation as smoothly as possible, whenever an opportunity arises; or (2) to backtrack openly: "I am sorry, but I forgot to mention . . ." The latter certainly leads to a flaw in the presentation, like a scratch on a brand-new car. The irritation to both speaker and audience will perhaps teach the speaker to be better prepared next time!

Formulation and Argumentation

"Is there a method to your madness?" is a question that sometimes appears in the eyes of the confused listener. To keep an audience motivated to invest energy in trying to comprehend your arguments, **it is essential that the lecture proceed as a logical unfolding of information. During the presentation facts must be enumerated in sequential steps, each step firmly founded on the previous one.** Every statement that provides new information should tie into information from the previous sentence. Since every sentence has the ability to alter the context or the interpretation of the presentation, explicit ongoing reference to the subject of the story limits the number of possible interpretations.

Consider, for example, a presentation that starts with the following statement: "In 1987, a team of British scientists sailed to the Galápagos Islands to study turtles. When they arrived, they were in poor condition." It is not clear to the audience what to expect next. Were the turtles in poor condition? Were the scientists in poor condition? Or were the islands in poor condition? Of course, the title of the seminar helps shape the context for the audience, unless the title is also uninformative, e.g. "Nineteen Eighty-Seven: A Memorable Year for a Group of British Scientists." The title "Turtles of the Galápagos Islands, An Endangered Species" would suggest that the speaker meant to indicate that the *animals* were in poor condition when the scientists arrived. Had the title been "Aftereffects of Seasickness in the Tropics," the audience might expect that the *British scientists* were in poor condition after sailing to the islands. Finally, "Erosion of the Shore Line of the Galápagos Islands" would indicate that the scientists, who had come to study the turtle population, noticed that the *islands* were in poor condition. Revised, such a statement eliminates ambiguity and prepares the audience for a presentation on the fate of turtles of the Galápagos Islands: "In 1987, a team of British scientists sailed to the Galápagos Islands to study turtles. When they arrived, the animals were in poor condition." Frequent statements of the explicit subject under discussion make it easier for the audience to follow the speaker's line of thought as the story unfolds.

It is useful at this point to introduce the concept of Umwelt. In ethology the German word *Umwelt* is defined as the total characteristic perception with which an organism perceives its surroundings, depending on the sensory organs with which it is endowed. Although operating in the same milieu, your audiovisual Umwelt is different from the predominantly olfactory Umwelt of your dog. Speakers each live in an intellectual Umwelt, a realm of concepts and insights with which they are so familiar that it has become second nature to them. Each member of the audience also has his or her own intellectual Umwelt, and it is where these Umwelts overlap that we can effectively exchange information.

Remember that your labyrinth of knowledge, with its familiar shortcuts, alternative routes, and interconnections is unfamiliar to the audience listening to your story for the first time. The audience has to be led by the hand through this maze of information to arrive safely at the treasure in the center. Many speakers when presenting experimental data presume that the audience is familiar with technical details which to them represent an everyday routine. They treat the experimental design as merely incidental and expect that the audience can fill in the gaps if they simply mention the conclusions of their data. For example, a speaker may present data as follows: "This Northern blot shows that this receptor is only expressed in the heart." The audience stares at the picture of the gel without really understanding what the blobs represent. It would be very little trouble for the speaker to modify his presentation as follows: "To investigate in which tissues this receptor occurs, we extracted messenger RNA from several different tissues, fractionated these RNAs on a gel, and immobilized them to a nitrocellulose membrane. We then probed the membrane with DNA that encodes the receptor in order to identify bands containing messenger RNA complementary to this DNA. As you can see on this slide, the only tissue that shows a band is from the heart. Thus, we can conclude that this receptor is only expressed in the heart." The speaker has now guided the audience through the data so that they can better understand and appreciate the results. Furthermore, explicit descriptions of experimental design display a speaker's knowledge and authority.

To communicate effectively, avoid the use of hyperbole or jargon whenever possible! Few members of an uninitiated audience may understand what is really meant by "the functional significance of defining the molecular parameters of a protein's structure;" instead of using complex terminology, simply ask how the protein's structure enables it to perform a given function. Similarly, when the molecular biologist talks about "footprints in the CAAT box," few members of the audience will recognize the reference to gel patterns that reveal a well-known regulatory region controlling the expression of many genes.

Scientists immersed daily in a world of professional terminology may not appreciate that what in their intellectual Umwelt are considered ordinary phrases are perceived as unintelligible jargon by an uninitiated audience. If the audience has to invest part of its limited energy to decipher unfamiliar terms, less energy will remain for addressing the scientific content of the presentation. In general, it is difficult for members of an audience to absorb new information and at the same time manipulate that information mentally and analyze it. They need all the help the speaker can give them to keep track of the story as it unfolds. It is easy to lose an audience, and once that happens it is usually impossible to reactivate their interest.

Precision in formulating your arguments is another prerequisite to delivering a strong scientific presentation. **Speech reflects our thought processes, and an imprecise speaker is often an unfocused thinker.** Moreover, unlike a written article that can be edited and revised, read and reread, an oral presentation gives the author only one chance to present his or her views. Once the words are uttered, they move into the audience's memory and outside the speaker's control. There is no revision, no editing, no correction after the fact. A glib, nonchalant speaker leaves a poor impression on the audience and can get in trouble during the discussion session if forced to retract or clarify imprecise or poorly formulated statements that were open invitations to attack. "That is not what I meant" is a very poor answer to a question that arises because the speaker was initially careless in formulating a point. A careful and thoughtful speaker impresses the audience as a solid scientist. There are big differences among the phrases "These data *demonstrate* that—",

"These data *support the notion* that—", "These data *consolidate* the notion that—", and "these data *suggest that*—." There is also a big difference between statements that something "*can* occur" and "*might* occur." New scientific ideas are often developed through combinations of hypotheses, speculations, and experimental evidence. **You should carefully analyze the often fuzzy borders that separate experimental evidence from speculation. The care with which this intellectual process is performed is reflected in the manner in which you formulate your sentences.**

Discussion and disagreement are two characteristic properties of the scientific endeavor. Frequently, we find ourselves compelled to present experimental data or a personal viewpoint not in line with generally accepted ideas or in conflict with data or hypotheses promulgated by a competitor. In these situations you have two options. You can ignore the prevailing ideas or the competing opinion and simply present your own observations, or you can discuss the points of dissent explicitly. The latter is almost a necessity if your idea is a radical departure from accepted notions rather than a minor disagreement, or if the dissenting competitor is a member of the audience. In such cases, **present the**

generally accepted viewpoint or the competitor's hypothesis first and subsequently formulate the arguments supporting your own point of view. This allows the story to evolve from the generally accepted viewpoint toward your own hypothesis while you present arguments to convince the audience that the straw man initially described is inferior to your newly developed model. In addition, your hypothesis — treated last — receives greater emphasis. To offer your own hypothesis first and then mention possible alternatives is likely to generate skepticism toward your ideas and place more emphasis on their limitations and alternatives. **As speaker, you have the opportunity to have the last word. That opportunity should not be ignored.**

Always be tactful and discreet when referring to data or hypotheses from other investigators that are in disagreement with your own ideas. Even if certain notions are considered simplistic or if some data are likely to be artifacts, you should never bluntly point this out to your audience. There are elegant and gracious ways to convey the same message. Begin by simply stating the ideas with which you disagree: "Dr. Jones obtained the following data, which illustrated that this agent has potent antihypertensive effects; based on these data, she hypothesized that these effects may be mediated via a target site in the central nervous system." Your description of her approach can now, in subtle ways, point out that the initial experiments by Dr. Jones were seriously flawed: "Since Dr. Jones's experiments were primarily done on old, pregnant, female rats, we decided to investigate whether similar effects could also be observed in male rats or nonpregnant females. Furthermore, rather than using the chemical measurements described by Dr. Jones, we used a highly sensitive radioimmunoassay, which shows absolute specificity for this compound. Our data, demonstrated on the next slide, show that the target sites for this antihypertensive agent are localized specifically to the kidney. Of course, this does not mean that in older, pregnant females the compound's target sites could not be located in the brain." Nobody in the audience will think it feasible that this drug suddenly affects areas in the brain of old, pregnant rats but not other rats. It is clear to the audience that Dr. Jones's chemical measurements were most likely flawed and that the sources of her animals were dubious. Your listeners will be convinced that your results are correct, even if Dr. Jones's original paper has been a classic for many years. If you boldly state that Dr. Jones is incompetent and the experiments artifactual, the audience will be shocked and Dr. Jones offended. **There is a difference between a street urchin's brawl and a gentleman's duel. The scientist should always fight like a gentleman.** (And of course, one never knows; in the end Dr. Jones may be right.)

To present sound and convincing argumentation throughout the presentation, always reflect on the questions: How solid are my data? Is it justified to draw these conclusions based on the facts that I am presenting? Are there alternative interpretations of my data? A critical assessment of your own presentation is essential to gain the listeners' credibility. Especially in science it is important to understand relationships of cause and effect. Most flawed arguments result from lack of appreciation for the difference between causation and correlation. For example, for many years the American Cancer Society has argued that smoking causes lung cancer, based mostly on the substantially higher incidence of lung cancer among smokers. Protagonists of the tobacco industry have challenged this contention as a mere correlation that fails to establish a direct, causative link between this disease and the activity of smoking. The migratory behavior of birds offers another example of the difference between correlation and causation. Every fall Canada geese travel many miles from their northern breeding grounds to the central and southern United States. Since the onset of their journey correlates with declining temperatures, one could conclude that colder weather triggers migration. However, it has been established that the shorter photoperiod (lengthening nights) rather than the cooler temperature directs the onset of migration. If different observations are made under the same experimental conditions, it is difficult to conclude that one of these observations leads to the other without showing this *directly and unambiguously.* **By recognizing the limitations of your experiments and clearly defining the conditions under which your conclusions are valid, you gain the respect and credibility of your audience.**

The issue of argumentation is also directly related to the amount of detail that should be included in your presentation. **Ask the question: What argument do I want to make? Present only information directly relevant to this argument. Peripheral, "decorative" information should be deleted.** When I advise one of my students to omit a slide from an upcoming presentation, I often hear "But it is such a pretty picture!" Pretty is not enough. The information needs to function within the presentation; if it has no functional merit, it becomes distracting and interferes with the main focus. On the other hand, the arguments that support an important point of the presentation should be complete. The audience cannot guess what type of supportive data the speaker is *not* presenting. Like a jury in a courtroom, it can only weigh the validity of the speaker's argument based on the data presented. A court indeed it is! The speaker is being judged, and the punishment for a guilty verdict—a seminar deemed mediocre—may be as light as a blemish on the speaker's reputation or as severe as the loss of a job opportunity or the possibility of funding.

The Conclusion: Brief and to the Point

Just as zooming in is essential in the early stages of a presentation to put the talk into perspective, **zooming out can be a valuable tool near the end of a presentation, when we can remind the audience once again that the data**

we have offered relate back to the major scientific principle with which we started. This, again, enables the audience to appreciate the significance of the work within a larger context.

Speakers frequently finish their presentations by acknowledging contributions by their colleagues and coworkers, often showing their names or pictures in the final slides. I prefer to write the names of coworkers or colleagues on the board and refer to them as I progress through my talk, pointing out the key names at appropriate moments in my presentation. Alternatively, I sometimes show the names of my contributors on slides just before I discuss their contributions to the overall work: "At that moment, Dr. X [slide with the name or picture appears] in my lab decided to follow up on these experiments and designed the following approach [next slide]." These strategies not only give more effective and explicit credit than simply listing all the various contributors at the end of the presentation; they also prevent the credits from diluting your final take-home message.

The conclusion of a scientific presentation should be firm and decisive, like the final chords of a Beethoven symphony. I have seen many potentially excellent seminars fizzle out or fade away because the speaker did not know how to end. **The conclusion of the presentation is its most important moment. It provides the take-home message, often the only thing that will be remembered. It determines the final impression and impact that you will make on your audience. The conclusion should always be reduced to a concise statement, preferably shown as text or a simple diagram on a slide or overhead transparency.** Sometimes speakers finish their presentations with long lists of conclusions that occupy several slides (often such a presentation is unfocused in the first place). It is impossible for an audience to absorb a diverse array of conclusions. **The conclusion should consist of a single major statement, with not more than two or three connotations, if these are absolutely essential.**

After stating the conclusion, the speaker should simply say "thank you" and stop talking. Any further words will distract from the conclusion and harm the presentation. **The conclusion should very clearly demarcate the end of the lecture.** I have listened to many speakers who show a clear and crisp conclusion and then feel compelled to continue to discuss their ideas for future research or to indicate what else their labs are doing. When the audience thought that the presentation was over, they were forced to listen to more, unrelated material for which they were not properly prepared. They could only guess when the talk would end. They started to look at their watches, and some sheepishly sneaked out of the room when the speaker turned toward the screen. As established earlier, the audience comes to a seminar equipped with a defined amount of listener energy. This energy is dispensed as the presentation proceeds and is exhausted at the moment when the seminar is expected to terminate. If your presentation goes only 5 minutes overtime, those 5 minutes represent less than 10 percent of the total allotted time of a 1-hour seminar. Yet the audience experiences this increment as disproportionately longer, and its attention drops dramatically. A car with an empty gas tank will not go that one extra mile, even though it has been driven all the way across town. A speaker who does not know when to finish can do irreparable damage to an otherwise excellent presentation. Violating the time limit generates in the audience the same feeling we have when stuck in a traffic jam on the highway; impatience and irritation mount every minute. **The most important rule for a scien-**

tific presentation is to finish on time and on a clear and resonant note. A single gunshot, one glamorous crescendo, one majestic moment—then silence.

We can now summarize four of the basic principles that enable us to structure a focused and coherent presentation, as discussed above (see also Figure 2).

1. **Three devices can put a presentation in the desired perspective:**

   a. Indicate the scope of the presentation by an informative title.

   b. "Zoom in" to the topic during the introductory segment of the presentation and "zoom out" near its end.

   c. Decide on the underlying question that the presentation seeks to address; then divide that question into a hierarchically organized array of subquestions, and develop the presentation as a series of answers to these questions.

2. **The mainstream of the presentation should address a single focal issue, tuned to the interests of the audience. Sidetracks from this mainstream should be brief and should always return to the same point in the main-**

stream where they started. Omit information not directly relevant to the focus of the presentation, and avoid backtracking.

3. The statements constituting the mainstream of the presentation should delineate a clear, logical line of thought. Formulate explanations of scientific concepts and experimental methodology unambiguously, without using professional jargon.

4. The presentation should end with a clearly formulated, concise conclusion. When the take-home message has been delivered, stop.

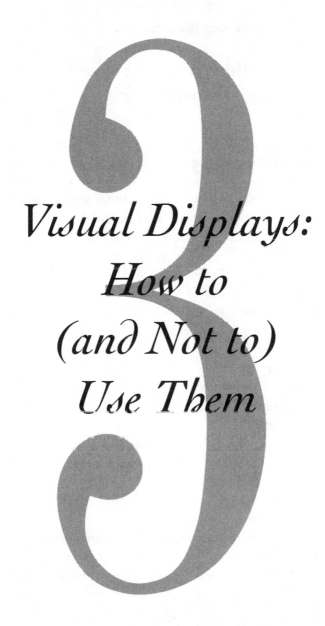

Visual Displays:
How to
(and Not to)
Use Them

Slides and Overhead Transparencies

Visual displays are at the very heart of scientific presentations, since they show the data. Appropriately chosen and well-designed diagrams, figures, and cartoons go a long way toward making a presentation successful. Since slides and overhead transparencies are still the most common means of data presentation, we will first consider their design and use. Later on we will cover the use of videodiscs, which provide not only high-quality visual displays but also full audiovisual integration. Videodiscs contribute convenience and flair, but the basic guidelines for the appropriate design of visual displays are the same whether the image is stored on an optical disc or in the form of a slide or transparency.

As indicated earlier, the pacing of slides should be reasonable, approximately 20 to 25 slides spaced fairly evenly for a 45-minute presentation. Slides have the advantage over overhead transparencies of allowing for a smoother flow of the presentation and easier transitions from one slide to the next. They also tend to be aesthetically more pleasing. Overheads have the advantage of allowing the speaker to add to the projected image during the presentation by writing directly on the transparency. Personally I find it distracting to watch the speaker's magnified hand block part of the image on the screen, while the speaker looks down at the overhead projector rather than maintaining eye contact with the audience. I prefer to watch a speaker use the blackboard when he or she wants to communicate ideas in writing during the presentation. However, transparencies come in handy when additions need to be made to a pre-existing complex figure during the talk. They remain popular especially among mathematicians and engineers, who like to write down information as they speak. Overhead transparencies also are considerably cheaper than slides and can be prepared at the last minute before a presentation. Nothing, of course, excludes the use of both slides and overheads in the same presentation.

When using slides, carefully double-check their sequence and orientation before the presentation. Nothing is more disruptive to the momentum of a seminar than an inverted or out-of-sequence slide that must be rearranged. Slides marked with an orientation dot on the frame in the lower left-hand corner will be projected correctly when each slide is placed in the carousel with the orientation dot in the upper left-hand corner facing away from the screen toward the back of the projector. If possible, bring your own carousel; you can arrange and double-check your slides at leisure well before the presentation.

Attention also needs to be paid to the illumination in the room. If dimmers are available, the room should not be darkened more than necessary to show the slides clearly. **Darkness invites the audience to daydream or fall asleep.** When the seminar contains a period in which slides are not shown, the projector should be turned off and the lights up. A slide that remains on the screen although the speaker has long since finished describing it, or an illuminated blank screen, becomes a distraction. Often slide projectors contain a noisy fan; turning it off for a minute may in itself be a welcome break. **Changing the illumination in the room according to ongoing needs also helps keep the audience alert and attentive.**

The three most important prerequisites for slides are that they be *clean, simple,* and *necessary in the story line.* Slides with smudges, stains, imperfectly drawn figures, or typographical errors reflect carelessness and sloppiness on the part of the speaker. Slides that have obviously been prepared in a rush and without precision are omens of a poor presentation. Clean slides are especially important for job interviews, where, like the speaker's attire, they are likely to be interpreted as signs of whether the speaker has lax work habits or is conscientious with an eye for detail and aesthetics.

Slides or overhead transparencies should illustrate a single point and, like the presentation itself, have only one focus. Images shown during an oral presentation differ from figures in a written document. They serve a different purpose: to communicate concepts and data to a listening and viewing audience. It is, therefore, often impossible simply to take an unmodified figure from a paper and present it as a slide without simplifying it visually and elaborating it verbally with a title and perhaps a brief summary statement underneath. Many journals limit the number of figures per publication, and as a result authors often compose elaborate panels. For a written publication this may be an effective way of showing as many facts as possible in the most compact and concise manner. During an oral presentation, however, the audience has only a limited opportunity to examine the data, while at the same time focusing its attention on the speaker. **Complex data delivered during a seminar cannot be fully appreciated unless the speaker separates them into a series of simplified constituents. Figures composed of multiple panels should be avoided** (Figure 3). **Instead, the individual panels should be presented sequentially as separate slides** (Figure 4 a and b). After showing the individual components separately, you can display the composite to demonstrate interrelationships or comparisons between panels. Similarly, **try to avoid showing tables.** Tables containing rows or columns of numbers are an excellent way to document data in written form, but nobody in the audience can read, compare, and analyze tabulated data points during an oral presentation (Figure 5). Instead, the data should be converted into a bar graph (Figure 6) or if possible, a line drawing (Figure 7) — both far more effective vehicles for conveying information during lectures.

Equations frequently appear on slides when the content of the presentation is mathematical or biophysical. For audiences consisting mostly of mathematicians, biophysicists, or engineers, equations present no problem. However, **most audiences find equations intimidating and are**

(Text continues on page 103)

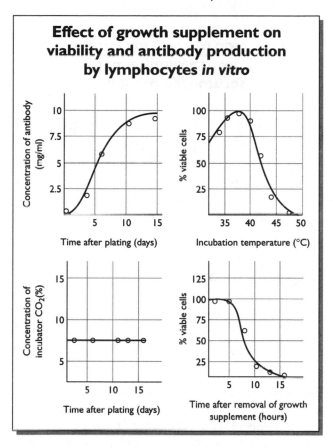

Figure 3 A large number of panels on a slide can interfere with effective presentation of information. As the speaker discusses one of the panels, the audience's attention is distracted by neighboring panels. Delete panels irrelevant to the presentation; offer relevant panels sequentially, as shown in Figures 4a and b. (The data presented in these figures are fictitious.)

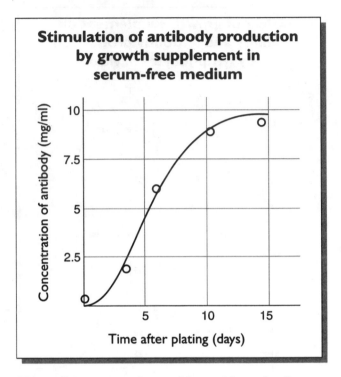

Figure 4 Present each set of data without the distracting presence of other panels. If it is necessary to compare two sets of data, they can be shown on one slide side-by-side as two panels, but only after the speaker has familiarized the audience with each set individually. (a) Graph 1

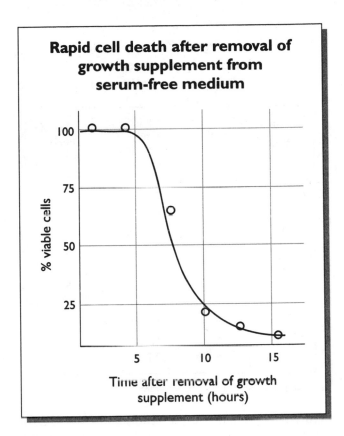

4 (b) Graph 2

Effect of growth supplement on antibody production by lymphocytes in cell culture

Time after plating (days)	Growth supplement	Antibody concentration (mg/ml) ± SEM	Number of cultures tested
1	+	0.1 ± 0.1	5
	–	0.2 ± 0.1	5
3	+	3.0 ± 0.4	4
	–	0.8 ± 0.1	4
5	+	12.0 ± 0.8	6
	–	2.5 ± 0.3	5
9	+	10.8 ± 0.4	6
	–	1.9 ± 0.2	5
14	+	11.7 ± 0.8	4
	–	3.2 ± 0.4	5

Figure 5 Information presented as a table. It is difficult for the audience to evaluate the effect of growth supplement on antibody production during the short time available to view this slide. The standard errors and number of cultures tested may be important in a written article, but are distracting in an oral presentation. Compare the same information presented in graphic form in Figures 6 and 7 (again, the data presented here are fictitious).

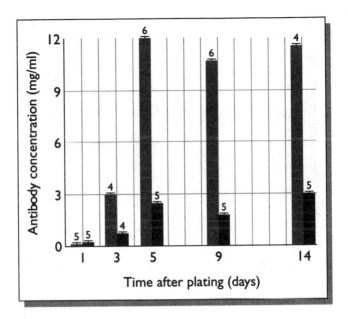

Figure 6 The same data as in Figure 5, presented as bar graphs. If the speaker deems standard errors and number of measurements important, bar graphs allow the relatively unobtrusive indication of this information above the bars and are often preferable to line graphs, where such statistical information may crowd the figure (see Figure 7).

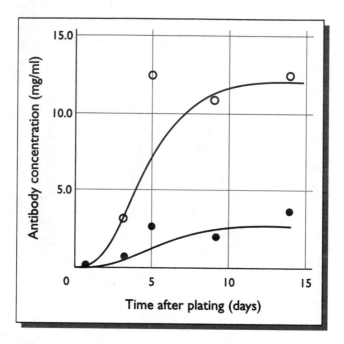

Figure 7 The same data presented in Figure 5, with open symbols representing the presence of growth supplement and closed symbols indicating the controls. Note that the effect of growth supplement is immediately evident in this representation. Omission of standard errors and number of measurements (which can be supplied verbally during the presentation) helps generate a simpler, clearer picture.

likely to "tune out" as soon as one appears on the screen. Often the significance of an equation can be described verbally (for example, "the increase in volume is proportional to the surface area of the cell").

Graphs should contain clearly labeled axes (Figure 8**).** Some precision may be sacrificed for the sake of brevity and simplicity. For example, the label "Adenylate cyclase activity $(\text{nmoles.min}^{-1}.\text{mg protein}^{-1})$" can in some cases, when presented on a slide, be replaced by the simpler indication "Generation of cyclic AMP." Lettering along the axes should be as large as possible. Only the minimum number of scale divisions should be indicated along the axes. When the Y-axis designates percentage of activity, for example, often only the 50th and 100th percentiles need to be labeled (rather than every 10th percentile). **The less busy a figure appears, the more justice it does to the information it attempts to communicate.**

I often see speakers present slides that are essentially invisible, a small figure with microscopic lettering lost in a vast ocean of light (Figure 9). A diagram on a slide cannot be too large. This is especially true for text. Make sure that your slides are designed so that the image fills the screen and that labels are large. One of the

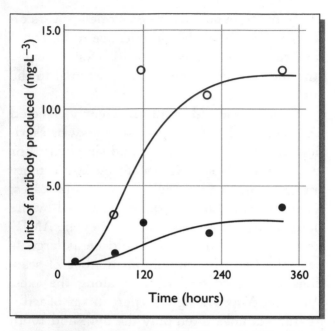

Figure 8 The same graph as in Figure 7. Using hours rather than days along the time axis and using the notation mg • L^{-3} rather than the simpler notation of mg/ml to indicate the antibody concentration make the graph harder to read. Adding unnecessary complexity to figures can cause an audience to "tune out."

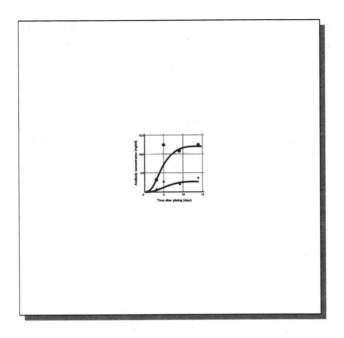

Figure 9 A smaller version of the graph shown in Figure 7 is lost in the center of the bright screen. This graph, essentially invisible to the audience, will not be understood as well as its larger counterpart (Figure 7).

most important rules that I impress on my students is: **LETTERING ON SLIDES CAN NEVER BE TOO BIG!**

When presenting text, some speakers like to use color contrast, such as white on blue or yellow on black lettering. This is, of course, a matter of taste, although personally I prefer black and white for clearest contrast. In any case, maintain uniformity among all the slides used in the presentation. **Uniformity of style throughout the presentation accentuates and underscores the flow and coherence of the talk.** Once a speaker has decided on a preferred style, it is easy enough to ensure that a collection of slides is built up in a uniform style.

Include only slides that are necessary to and functional within the story line. Since the quintessential purpose of slides is to communicate information effectively, delete all information that is irrelevant to the presentation. When pictures reproduced from published articles are shown, any text, including the original legend, should be cropped away so as not to distract from the figure (Figure 10). **However, a slide can be enhanced by adding a title to the diagram (Figure 11) and a line or two that concisely describe the conclusions to be drawn from the slide (Figure 12). Credit to**

ions at 1-week in-
der anesthesia, and
the 45% ammonium
alin fraction. After
sive dialysis against
NaCl, and 10 mM
reakdown products
d be absorbed on
urified by affinity
to agarose (Sigma
f Goudswaard et al

ibody competition
b-well Immulon II
nc., Chantilly, VA).
nspension of cillary
e buffer, pH 9.6, at
ne plates were incu-
y with PBS/Tween,
ate buffer, 100 mM
20, and 0.2% (w/v)
plates were washed
wells were incubated
desired dilution in
ces were washed as
nonium sulfate cuts

color formation reflecting bound mAb was measu
nm in an ELISA microtiter plate reader. Maxim
of mAb was evaluated by substituting preimmune
the polyclonal antiserum in the competition assay
of the rabbit antiserum was verified by deleting s
incubation with mAb, visualizing bound antibody

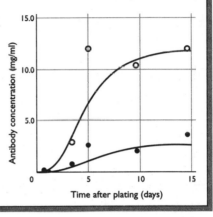

Figure 10 The graph shown in Figure 7 as if taken straight from a publication. Note how the text around the graph distracts the audience.

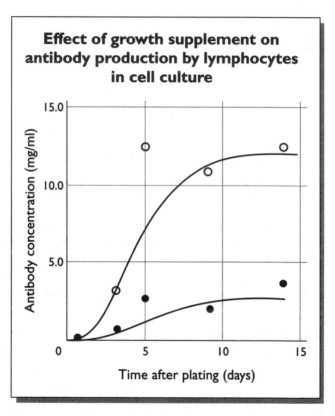

Figure 11 The graph shown in Figure 7 has been enhanced by a title, which makes its purpose immediately clear to the audience.

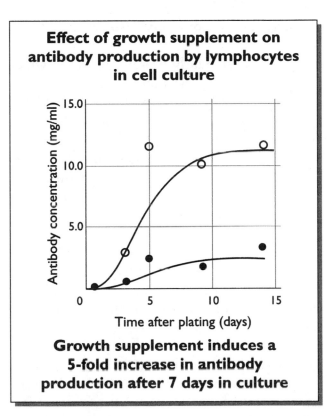

Figure 12 The graph shown in Figure 7 has been further enhanced by a brief summary statement. The message that the figure conveys should now be crystal clear even to the least attentive member of the audience. Compare the table presented in Figure 5 or the original graph in Figure 7. Note, however, that too much text may make the figure too busy. How much text to include is always a judgment call.

the original authors may also be added to a slide.

Speakers regularly present images in which crucial information is poorly depicted. Sometimes the speaker is aware of this and offers a sheepish apology — "You may not be able to see this, but take my word for it. . . ." This is embarrassing and creates a subconscious feeling of skepticism. It is better simply to mention the information without showing a dubious slide. Examples of important information often presented in an essentially incomprehensible manner are amino acid sequences of proteins or nucleotide sequences of DNA. The myriad small letters on the screen convey no significant information (Figure 13). Such sequences are better displayed diagrammatically (Figure 14) or as horizontal bars in which parts of the sequence can be highlighted (Figure 15). A series of bars representing a family of related species can contain colored or shaded boxes to indicate regions of homology. Similar visual codes can be used to designate transmembrane domains of proteins, leader peptides, sites that control initiation of transcription, and a host of other properties. Representing protein or DNA sequences in this manner accentuates those aspects of the sequence that constitute important

Amino Acid Sequence of an Odorant Receptor

MTEENQTVISQFLLLFLPIPSEHQ<u>HVFYALFLSMYL</u>
                                     I

<u>TTVLGNIIIILIHL</u>DSHLHT<u>PMYLFLSNLSFSDLCFS</u>
                              II

SVT<u>MP</u>KLLQNMQSQVPSIPFAGCLT<u>QLYFYLYFAD</u>

<u>LESFLLVAMAY</u>DRYVAICFPLHYMSIMSPK<u>LCVSLVV</u>
  III

<u>LSWVLTTFHAML</u>HTLLMARLSFCADNMIPHFFC
  IV

DISPLLKLSCSDTHVNE<u>LVIFVMGGLVIVIPFVLIIV</u>
                             V

<u>SY</u>ARVVASILKVPSVRGIHK<u>IFSTCGSHLSVVSLFYG</u>
                               VI

<u>TIIGLYL</u>CPSANNSTVKE<u>TVMAMMYTVVTPMLN</u>
                             VII

<u>PFIYSL</u>RNRDMKEALIRVLCKKKITFCL

Figure 13 This figure shows the amino acid sequence of an odorant receptor. Proposed transmembrane regions are indicated by roman numerals and underlined. The audience experiences this slide as white noise—meaningless letters. Compare to Figure 14.

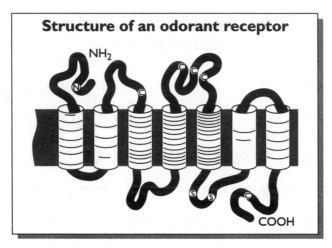

Figure 14 The same odorant receptor as in Figure 13 is here represented as a diagram showing the transmembrane orientation of the polypeptide. Not only has the figure eliminated the white noise, but it displays more information, such as the extracellular localization of the amino (NH_2) end as opposed to the intracellular carboxyl (COOH) end. The extent of shading of the transmembrane domains reflects the extent of variability in sequence observed in these domains among different odorant receptors. Only those amino acid residues that may be important for the attachment of sugars (N), phosphorylation (S), or protein folding (C) have been indicated. Compare this representation of the odorant receptor to that in Figure 13. From a scientific point of view both figures contain the same information, but only this one can communicate effectively to a viewing and listening audience.

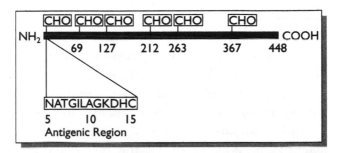

Figure 15 This schematic representation of the amino acid sequence of a protein as a horizontal bar indicates only information salient to the presentation—in this case, the relative positions of sugar groups (CHO) and an antigenic region that the speaker will discuss in detail.

information, while deleting an enormous amount of irrelevant documentation.

The presentation of an extended sequence of information clearly illustrates yet another great challenge of scientific presentations: to convey a number of complex concepts in a readily understandable manner. To achieve this, it is essential initially to break the concept down into an array of simplified components. Two strategies meet this challenge through the creative use of slides or transparencies. **One way to simplify complicated diagrams is to use a sequence of slides showing the same concept or process, but with increasing complexity** (Figure 16 versus Figure 17 a and b). For example, when discussing neuronal circuitry in an area of the brain, you can first show a slide that illustrates the primary pathway—two neurons synapsing onto each other. The next slide can offer the same image, but add to it excitatory interneurons, which contact the first two cells. The final slide of the series can show the previous image with the addition of inhibitory interneurons. If the last slide had been shown without the first two, it is likely that the audience, baffled by the complexity of the neural connectivity, would have lost interest. Electrical wiring diagrams provide another example of

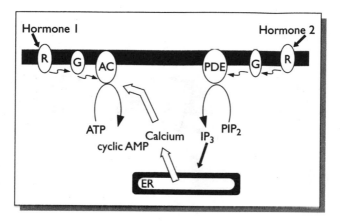

Figure 16 Too complex a diagram may overwhelm an audience who can view it for only one or two minutes. This diagram represents two different metabolic pathways activated by two different hormones. "R" designates receptors for these hormones and "G" designates G–proteins that link the activation of the receptors to stimulation of enzymes that generate intracellular second messengers. The pathway on the left activates adenylate cyclase (AC), which produces cyclic AMP; the other pathway activates a phosphodiesterase (PDE), which liberates inositol triphosphate (IP$_3$) from phosphatidyl inositol bisphosphate (PIP$_2$). The liberated IP$_3$ then causes the release of calcium from cisternae of the endoplasmic reticulum (ER). Calcium resulting from activation by hormone 2 regulates the adenylate cyclase controlled by hormone 1. The same scheme presented as a sequence of slides (Figures 17a and b) builds up the complexity gradually.

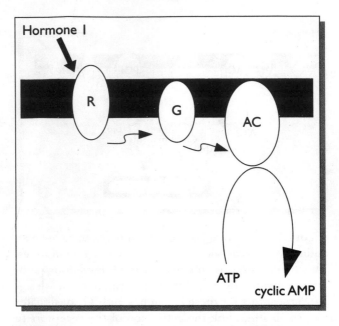

Figure 17 (a) First the pathway activated by hormone 1 is shown above.

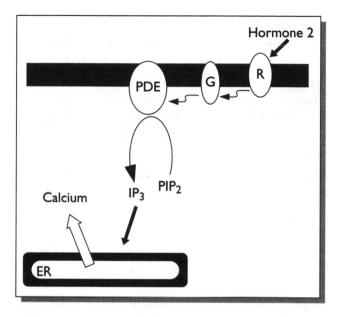

Figure 17 (b) Then the speaker introduces the pathway activated by hormone 2. Now the speaker can present the complex scheme of Figure 16, having first familiarized the audience with its components.

complex representations that can be difficult to appreciate on a single viewing. Such diagrams can be rendered more palatable by first showing a simplified version of the basic connections, and gradually increasing the complexity by adding components in a series of sequential slides. The audience, seeing the diagram created step-by-step, will become familiar with its different components; they will no long-er feel overwhelmed when the complete circuit is displayed. By building up a complex diagram from its components and allowing the audience to follow its development, the speaker retains the audience's attentiveness and appreciation.

The second way in which complex issues can be simplified is through the creative use of color. Color-coding components of an image and consistently maintaining this code through a series of slides facilitates the audience's understanding. One of my students once lectured on the cellular architecture of the retina. He first showed only the primary neural pathway—the connections between retinal rods and ganglion cells and the projections of the latter to the central nervous system. These cells were all colored red. In the next slide the same image appeared, but amacrine cells had been added and were colored green. His last slide showed the complete circuitry of the retina and its projections to the central nervous system, using three different color codes. It was one of the best explanations of the neural circuitry of the retina I have ever seen.

Transparencies can build up complexity in yet another way, by superimposition: components can be added to an image simply by laying another transparency on top of the first one. One speaker made creative use of both color and superimposition to describe the fate of migrating cells during embryogenesis. Superimposition of transparencies allowed her to gradually add complexity, as she described the ontogenetic process. At the same time, color-coding the cells enabled the audience to follow with ease as each different cell type reached its final destination. Transparencies allow attention to be focused on different components of the image by coloring them during the course of the presentation. Personally, I prefer to see figures neatly prepared in advance, since it gives a more organized aura to the presentation.

Of course, there are instances in which it is impossible to simplify a complex diagram. In such cases the best strategy is to inform the audience immediately — up front — of the conclusion of the diagram. Then proceed to de-

scribe the figure and explain how it leads to the stated conclusion. In this manner the audience at least will appreciate why the complex diagram is necessary and important, and they should be motivated to pay attention as the speaker describes the evidence contained in the figure that will convince them that the conclusion is justified. Frequent summary messages presented as text on slides can protect the speaker from losing his audience when discussing a complicated topic.

At the end of the presentation, a slide or transparency with a concluding statement is invaluable to emphasize the take-home message. Always keep summary statements concise and to the point. Conclusion slides should contain only essential information and should not be muddled by ancillary facts of secondary importance that distract from the main take-home message (Figure 18 a and b).

One rather irritating technique that should be avoided is to cover part of a transparency and gradually disclose more of the information as the talk progresses. This results in a dark, hidden area on the screen under or above the figure. The speaker may think that gradually revealing secrets may keep the audience's attention by appealing to their curiosity. This is true;

Conclusions

1. Lymphocytes grown in our serum-free medium produce modest amounts of antibody.

2. The addition of growth supplement to the medium stimulates the production of antibody.

3. The stimulation of antibody production by the addition of growth supplement to these lymphocyte cultures causes a production of antibody fivefold larger than that observed in the absence of growth supplement.

4. The production of antibody saturates after seven days in culture both in the presence and in the absence of growth supplement.

5. The increase in antibody production by this growth supplement may be important for future studies.

Figure 18 (a) A slide concluding a scientific presentation by summarizing the data presented in Figures 5–12. This slide contains too many conclusions and too much text for the audience to absorb. It fails to convey a brief, crisp take-home message.

Conclusion

Growth supplement causes a fivefold increase in antibody production by cultured lymphocytes.

Figure 18 (b) A revised version shows a clear take-home message.

121

however, the audience will focus primarily on the hidden part of the image to be revealed rather than on the part the speaker is trying to describe. This technique of using transparencies is, therefore, distracting and detrimental to a presentation. Better to go to the small expense of buying a few more transparencies that can be shown sequentially than to cramp several pieces of information onto one sheet and gradually divulge them.

Sometimes a speaker refers to a figure several times during a presentation—for instance when returning to the mainstream after a sidetrack, as discussed earlier. In such cases the diagram central to the presentation could be permanently displayed on the blackboard. This is not always possible, however, especially when the blackboard is obscured by the screen or if the figure is too complex to draw on the board in a way that does it justice. **If the same figure has to be shown more than once, use duplicate slides rather than disrupt the momentum of the presentation by turning the carousel in the projector back to a previous slide and afterward advancing it via a series of slides that have already been shown.** When using transparencies it is a good idea to use duplicates as well, although it is possible to lay a transparency aside for repeated use later in the talk. With a duplicate there is no need to engage in manipulations to keep the transparencies organized; you can dedicate full attention to the delivery of the talk.

The simultaneous use of two slide projectors (or one slide projector and one overhead projector) enables the speaker to keep one image on the screen for extended periods of time while the second projection beam continues the flow of the lecture. When using two projectors, it is essential that one machine project stationary reference images, replaced relatively infrequently, and the other carry the main sequence of the presentation. The first projector can also be used at appropriate moments to allow comparison between two images. Familiarize the audience with the stationary image before proceeding to show images on the second screen. To ensure that the second projector is the one that provides continuity to the presentation, the first projector should be turned off whenever it is not essential. Randomizing the appearance of images between the two projectors, moreover, should be avoided. Although the use of two projectors can add luster to a presentation, there are dangers inherent in the approach besides the initial organizational headache—namely, the potential to diffuse the attention of the audience. It is important to prevent the stationary image from distracting from the images that compose the mainstream of the story. Dual projectors are to be used only by seasoned speakers and are risky in the hands of a novice or an inexperienced projectionist.

Video, Audio, and Props

Dynamic processes can be illustrated through a series of images showing the same phenomenon at different points in time (Figure 19). During the last decade, however, sophisticated techniques have been developed to monitor or simulate processes as a function of time. Concomitantly, the use of video during scientific presentations has become increasingly more popular. Whereas in the old days unreliable film projectors had to be used to show often poorly prepared film sequences, modern VCRs allow the convenient presentation of high-quality videotaped information. In some cases, as in

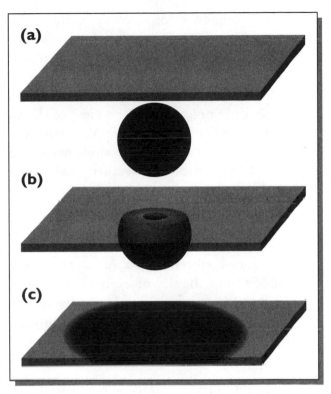

Figure 19 The fusion of a vesicle to a membrane can be shown through a series of images as an alternative to video. (a) The vesicle approaches the membrane. (b) Fusion occurs. (c) The vesicle has been incorporated into the membrane. These images could also be shown in sequence as separate slides. Notice how different shading allows the audience to keep track of that part of the membrane that previously constituted the vesicle.

time-lapsed sequences of migrating cells, videotaped evidence is indispensable for the maximum appreciation of scientific observations. It requires a screen that is readily visible from all positions in the lecture room, but which, when not in use, does not obstruct the view of slides, the blackboard, or the speaker. A system that can be wheeled into position when needed is optimal. Some modern lecture halls are equipped with a backstage projection system that allows video projection onto a glass screen from behind while enabling the projection of slides onto the front of the screen. Such technically advanced systems also often allow the video image to be the size of a cinema screen. It is advisable for speakers presenting video sequences or using multiple projectors to inquire about the visual display capabilities of the lecture hall well ahead of their scheduled presentations.

Video provides a wonderful tool for enlivening scientific presentations, but it has to be used prudently. Video sequences should be no longer than 1 minute; otherwise the videotape takes over the presentation and diverts attention from the speaker. For this reason the speaker should provide the commentary rather than having a sound track on the videotape it-self. Several short stretches of video intermingled with the slide presentation are guaranteed to keep the audience alert and attentive. **The speaker should be well organized and have a carefully edited tape to avoid fast-forwarding or rewinding during the seminar.** The speaker should also take some time before the seminar to become familiar with the video equipment provided, to insure smooth transitions between video and slide sequences. The use of videodiscs, described below, facilitates integration of video sequences with static images, since both are stored on the same disc.

Sound is not often used during scientific presentations, but like video it can arouse the audience's interest when used creatively. I have for the last several years given a lecture to the medical students at Duke University on cellular mechanisms of sensory transduction. I am usually scheduled as the second lecturer, with a brief break between my talk and the previous one. During this break I play Bach's *Goldberg Variations* over the sound system in the lecture hall. I am always delighted to see the surprised faces of the students when the first notes from the piano hesitantly emerge at the beginning of this beautiful piece. As I begin my lecture, I arrange for the music to fade out slowly while I tell the audience that Mother Nature, in designing cellular mechanisms for sensory transduction, used variations on a theme just as Johann Sebastian Bach did in composing his *Goldberg Variations*. At the end of the presentation the music starts again as the students pack up their belongings and leave the hall. It is always an extremely successful lecture, in which interest, curiosity, and attention are established even before I begin to speak. The music is used to underscore the take-home message of the lecture. Moreover, it provides an associative stimulus — many of the students are likely to remember my lecture whenever they listen to Bach and, especially, to the *Goldberg Variations*. I have also heard audio tapes effectively supplement lectures on echolocation in bats and on the development of bird song. The speakers brought cassettes with them so that their audiences could appreciate the actual sounds in addition to the visual representations.

Anything out of the ordinary usually gives a presentation that special memorable touch, setting it apart from others. We may think that "show-and-tell" is a childish technique that only applies in our elementary school days. This is not true. Show-and-tell is a teaching method that remains effective throughout our lives. Whenever I talk about olfaction, I never fail to arouse the audience's interest by giving them the opportunity before the lecture to smell some sample odors that I bring along as a demonstration. When I discuss the structure of taste buds, I usually bring in a six-pack of soft drink cans. These are often used as models for taste cells with their small apical membranes (where the cans open) exposed on one side and separated from the large basolateral membrane (the rest of the cans) by tight junctions (the plastic filler, which holds the cans in the six-pack together).

I remember how excited we were as children when the biology teacher brought in the human skeleton to teach us about bones and muscles, or when we were shown a plastic model of the brain that could be taken apart to show its various constituents. Chemistry classes were fun because of the inevitable malodors and explosions. Props do not lose their magic as we grow older. It is still a lot of fun to put on the 3-D glasses handed out by a speaker who discusses depth perception or who demonstrates optical illusions that may help tease apart the mechanisms by which visual information is processed in the brain. I do not remember every seminar I have listened to, but I do recall most presentations that used video, audio, or props. Not only were they informative; they were fun—and after all, enjoyment and pleasure motivate us more than anything else to pay attention.

Laser Discs: Integrated Visual Displays for the Twenty-first Century

During the last decade the development of compact disc technology has revolutionized the sound recording industry. The ability to store digital information on laser disc, however, has also been exploited to store and project images. As we are about to enter the twenty-first century, videodiscs are likely to join conventional slides in many lecture halls. The ability to store soundtrack, static and dynamic images, and computer animations on the same device provides remarkable convenience for audiovisual integration of information displays. Although at most universities only some lecture halls are currently equipped for the use of laser discs, upgrading of facilities and an anticipated decline in the cost of multimedia equipment as it gains in popularity are likely to increase the future use of videodiscs. For the preparation of visual displays, many institutions already have sophisticated graphics facilities equipped with scanners, computer-interfaced slide makers, and advanced graphics programs. Availability of any computer with a laser disc drive allows slides and images created with or modified by such graphics programs to be stored directly on disc. Many commercial photography shops can also transfer existing photographs or slides onto videodisc.

Retrieval of information from the disc during the seminar is accomplished by remote control, which allows the selected image to be displayed directly on the projection screen. The remote control device can be linked to a computer, a bar code, or an index print chart that provides the speaker with a pictorial index of the images on the disc and allows projection of an image simply by touching its corresponding picture on the chart. Of course, the quality of the image will still depend to a large extent on the projection facilities in the room.

In addition to eliminating the embarrassment that results from wrongly oriented slides or slides that are out of sequence,

videodiscs provide two major advantages: (1) achievement of convenient audiovisual integration of images, video sequences, sound, and animations, and (2) ability to manipulate the projected image. Images stored on videodisc are randomly accessible and can be shown in any sequence, eliminating the need for duplicate slides or rewinding videotapes. The speaker has the flexibility to return to any image at any time during the presentation, to make last-minute decisions about which image to show, and to choose from among a vast number of images stored on the disc. Moreover, images stored on videodisc can be rotated and scanned by "panning"; in some systems, parts of the image can be zoomed up during the presentation by a simple remote command. Moreover, images stored on videodiscs can be accessed via any computer equipped with a compatible drive, and with appropriate software they can be continuously modified, upgraded, and captioned. Since optical discs now available can store thousands of images, they provide a convenient, space-saving library for the storage of visual information. Offering so many advantages over conventional slides, videodiscs are likely to become a popular method for audiovisual displays in the future.

Handouts

Handouts are valuable for two applications. They serve an instructional purpose when accompanying lectures, and can also provide material (*e.g.,* curriculum vitae, a description of a project's history) that can be evaluated subsequent to the presentation, if the talk is part of a job interview or a research proposal. Instructional handouts can vary in length, whereas handouts that accompany interviews or research proposals should always be brief and to the point.

It is often advantageous to make instructional handouts such as lecture notes available prior to the actual talk, to give the students an opportunity to familiarize themselves somewhat with the content of the "upcoming attraction." They will then be able to focus all of their attention on the lecturer without being distracted by the handout, and they can pay special attention to concepts they found difficult to understand just reading the notes. The lecture here serves to clarify and cement the information, and **the handout and lecture have become two equal, integrated partners in the educational process. The handout can contain a vast amount of** information that could not possibly be covered in the accompanying lecture. The lecturer's job is to make clear, in the presentation, which information is central and which is peripheral to understanding the material.

I once participated in a series of lectures given to high school teachers to familiarize them with recent advances in the neurosciences. It was clear that a handout to which they could refer later would be useful, but at the same time, few of the teachers would have the time and inclination to read many pages on any one of the numerous topics to which they were exposed during the series. Therefore, I designed a single-page handout. My lecture dealt with the olfactory system, and my handout contained the three most crucial images of my presentation: a schematic diagram of the histology of the olfactory neuroepithelium, a figure that clearly illustrated how different odor molecules elicit different patterns of neural activity in subpopulations of olfactory neurons, and a model of the transmembrane organization of a putative odorant receptor. Each diagram had a title, a short legend in smaller print, and one sentence that provided a brief take-home message. Although my presentation covered more information in greater detail than was presented in the handout, that page insured that the recipients would be able to refer at a glance to the most important concepts of my presentation.

The Old-Fashioned Blackboard

Technological advances in our modern projection capabilities have gradually diminished our respect for what is perhaps the most effective teaching tool of all: the blackboard. Few speakers use it at all, and those that do use it sparingly. Yet those of us who grew up with the old-fashioned blackboard vividly remember how we had to strain to keep up with the teacher who wrote equations or diagrams on the board, while we simultaneously tried to copy the precious information into our notebooks. One swipe with the eraser and the knowledge inscribed on the board would be lost to us forever! The blackboard required us to put effort into the process of learning and made us active participants in the classroom. The creation of diagrams or text in front of our eyes by a lecturer who moves around to write on the board, while addressing the audience, provides a more dynamic experience than simply seeing the complete image appear on the screen. Sure, the blackboard has its shortcomings. It is less fancy, less aesthetic, and less convenient than slides. More than any other visual aid, however, it commands active attention from the audience. It remains a powerful teaching tool.

The blackboard can be used effectively in combination with slides and handouts. An outline of the presentation on the board can help the audience keep track of the story as it unfolds. Similarly, a diagram central to the presentation, when drawn on the board, provides a reference point to which the audience can resort at any time. Some of my students have used the board to create a list of arguments supporting (or contradicting) a hypothesis as it is being discussed. Others have used it during the lecture to add complexity to initially simple diagrams placed on the board before the presentation. This is similar to the use of a series of slides or transparencies that show an increasingly complex image, as discussed earlier. Many speakers, I among them, like to write the names of their collaborators and coworkers on the board to acknowledge their participation in the project being presented.

Some projection screens lower in front of the board. **When using the blackboard in com-**

bination with slides or overhead transparencies, make sure that information written on the board before the presentation will remain visible when the projection screen is in use. Remember also to have the room lights turned up whenever attention should be focused on the board. You should ascertain before the presentation that chalk or, as in many cases nowadays, erasable markers are available in a variety of colors, along with an eraser and a pointer. Finally, remember that it is frustrating for an audience when the speaker stands in front of the board and obscures the information being discussed. This generates the same feeling we have when an uncle stands up in front of the television just as a touchdown is about to be scored.

The board can be a highly effective teaching tool in combination with handouts. One of my colleagues gave a superb lecture to medical students on the interconnections between the limbic system and other regions of the brain. The handout the students received contained an empty diagram showing the different areas of the central nervous system. The same diagram was drawn on the board, and as she filled in the different pathways in that diagram, so did the students on their handouts. **There is no better way to teach than to actively involve students in the process of acquiring knowledge.**

Any lecturer can give a presentation. It takes a scholar to use the blackboard effectively. Slides project information for us in its entirety, but to generate diagrams and text on the blackboard requires active thinking, intellectual discipline, and an organized mind. Maybe we respect the blackboard because the brain rather than the projector conjures up the information. Unlike the slide projector, the blackboard forces speakers to think on their feet during the delivery of the presentation. The board is most effective in small, intimate seminar rooms. Poor visibility may compromise its use in large lecture halls, where slides remain the best visual aid.

One well-known biophysicist refuses to use slides and only employs the blackboard. Throughout his presentation, he fills the board with equations, diagrams, and figures. He defends the idea that the blackboard is the only acceptable way to communicate with the audience on, as it were, a one-to-one basis, and to convey information effectively. Some of his students have adopted the same philosophy. Their lectures are always excellent and have the old-fashioned, scholarly aura of an intimate teacher-student relationship around them. Besides admiration, however, this Amishman scholar, who shuns the slide projector as a matter of principle, draws a lot of criticism. Many of his colleagues have expressed the desire to see some real data occasionally, rather than their improvised representations on the board. A counterargument might be that the actual data will ultimately be documented in the scientific literature, while the purpose of a lecture is to communicate the underlying ideas, approaches, and tenets. Still, a modern-day scientific audience will find a graph drawn by hand on the board less compelling than a slide that shows the actual data points with their standard deviations and appropriately labeled axes.

The old-fashioned blackboard with its chalk (or its modern counterpart, the white board with its erasable markers) still plays an important supporting role in the drama of scientific presentation. To use all available resources in moderation optimizes our achievement. This holds as true for the use of visual aids during a scientific presentation as for everything else in life.

Poster Presentation: The Young Scientist's Debut Performance

Poster sessions are often the forum of choice for the young scientist who ventures for the first time into the scientific arena at a local or national convention. Poster presentations provide an intimate, low-pressure, and nonthreatening opportunity to exchange information with colleagues in the field. In contrast to slide presentations, they allow direct personal contact, and as a result they enable not only the presentation of information, but also a true exchange of ideas. The chance for young scientists to receive instant feedback in the form of constructive comments, criticisms, and suggestions makes the poster session especially valuable. **To benefit most from a poster presentation, you should consider it primarily an opportunity for exchange of ideas and dialogue, rather than merely a forum for data presentation.**

A poster is a visual display, subject to the same guidelines for presenting data that were previously outlined for the preparation of slides and transparencies. **The poster should be aesthetic and clean.** Make sure that it does not exceed the dimensions of the poster board—

usual standard dimensions are 1.1m (3'8") × 1.75m (5'8"). The poster should not flop beyond the edges of the board, where it will interfere with neighboring posters or sloppily hang down below eye level. **When choosing posterboard, select a muted background tone** such as gray, beige, light blue, or white. Screaming colors like pink, orange, or purple focus attention on a poster, but distract from the actual information presented. Personally, I find it painful to try to view data against a psychedelic background. If for easy transport the poster is composed of several panels, attempt to keep them similar in size so that they fit together neatly to form the complete display. The relationship between components of a poster that are conceptually connected can be accentuated by placing them together against a background shade slightly different from that of the rest of the poster. Many institutions have the capability of producing full-size photographed posters that can be transported as scrolls and easily set up. These posters always look neat and tidy, but they provide less flexibility for last-minute revisions than posters organized on cardboard panels.

Figures and diagrams displayed on a poster should be designed to be viewed from a distance, with clear and legible graphics (Figure 20). Each figure should be clearly identified with a number or letter at least one inch in size. To provide coherence and focus, each image should make only a single point. Again, **simplicity is the key to success.** Like a slide presentation, a poster presentation differs from a publication in that the author is on site to explain the figures and diagrams. Details in legends and descriptions can, therefore, be kept to a minimum. A concise summary above or underneath each figure in large, bold type will enable the viewer to conveniently absorb the take-home message of every component of the poster. **Like any scientific presentation, the poster should *tell a story*. Choose a brief and informative title and provide, in the upper left-hand corner of the poster, a concise introduction that indicates why the work presented is important within the context of a major scientific principle. Describe the approach in an engaging, condensed style without excessive detail, and organize the presentation of data in a logical, coherent sequence. The lower right-hand corner of the poster should contain a small number of well-phrased conclusions and a major, concise summary statement.** Like any scientific presentation, the poster should focus on a single concept. I have seen many young scientists who, driven by an unbridled desire to display as many of their achievements as possible, cram into their posters a jumble of unrelated data. Usually this results in a disorganized, unfocused display, the significance of which is hard to appreciate. **Include only material relevant to the story line.**

Arrange the figures and diagrams in vertical columns rather than horizontal rows. It is easier for viewers to scan a poster when they are not required to zigzag back and forth in front of it. Vertical organization of data on the poster also allows different viewers to study the poster simultaneously and move along sequentially without bumping into one another.

The presenting author should be available for explanation at his poster at the designated time and wear his name tag for easy identification. It is a good idea to come equipped with extra thumbtacks, tape, or glue for emergency repairs. The presenting author should stand in a position that does not obscure the poster or interfere with traffic. Viewers should be allowed to look at the poster at their leisure without being pestered by overly aggressive or energetic authors.

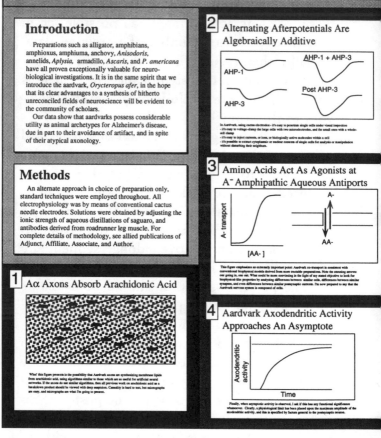

Figure 20 Sample poster showing suggested elements and lay-out, designed for the 1989 Annual Meeting of the Society for Neuroscience by Drs. Daniel Gardner of Cornell University

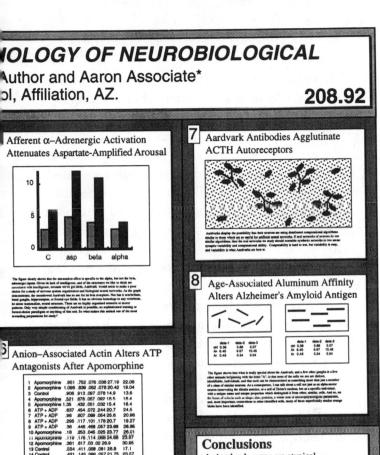

Medical College and Esther Gardner of NYU School of Medicine. Reproduced with permission.

It happens to me often at meetings that I accidentally glance at a poster while making my way to the restroom or cafeteria, only to be gratefully snared by the presenting author who, like a used-car salesman in the off season, desperately insists on providing me with a detailed tour of his display. Needless to say, I remember such an individual only as someone to avoid for the rest of my life. A presenting author at a poster session should behave like a waiter in a first-class restaurant, who is there when needed but does not aggravate the guests by interrupting conversation every ten minutes to inquire whether they are enjoying the food. The author should provide explanations when necessary and, above all, be open to discussion if he senses that his audience is willing and interested. Be prepared to provide a **brief** description of your poster when asked to do so. It is essential to keep such explanations short, since most visitors view a large number of posters one after the other and their attention span for each individual poster is limited. Also, other interested colleagues may be waiting in the wings. A well-organized poster will make it easy for a newcomer to enter immediately into an ongoing explanation, after briefly viewing the title and a few of the summary headings above the figures.

Remember that it is not the number of people who come to view your poster, but the quality of interactions with them that determines its success. Often a detailed informative discussion with a single interested and knowledgeable colleague can be more useful to the presenting author than continuously rattling off the explanation of the poster to crowds of casual viewers. Poster presentations provide a unique opportunity for young scientists to build relationships. Conversations in front of posters can lead to collaborations or establish long-term relationships that prove to be important career boosters. When you introduce yourself to a viewer who shows a clear interest in your poster, try to remember his name and perhaps write down his address. You may meet again on future occasions, and he will be pleased to find out that you remembered the discussion you had with him. Follow-up letters to important people in the field who came to view your poster, indicating that you enjoyed meeting them and discussing your data with them, go a long way to establishing valuable relationships and will provide them with a lasting memory of you and your work. Handing out reprints or copies of an abstract to your poster visitors will enable them to reflect on your data afterward and, among the multitude of posters viewed, will make your presentation memorable to them. One of my colleagues prepared small reproductions of his entire poster, which he sent to all of his poster visitors after the meeting. Receiving the reproduction of his poster in the mail allowed me to look at his data again in the quiet surroundings of my office and enhanced my appreciation for his work.

Finally, **science should be fun!** Anything out of the ordinary will attract attention and draw people to a poster they might otherwise ignore. There is an increasing trend among presenting authors to show video sequences along with the poster. With the advent of videodiscs, it is likely that in the future video monitors will play an increasingly important role in data presentation during poster sessions.

Imagination, however, requires no multimedia technology. A colleague of mine once presented a poster that examined the genetic predisposition of human beings to perceive the boar pheromone, androstenone, which has a strong odor to about half of the human population and is odorless to the rest. From his poster dangled a little sniffing device consisting of a tube that held cotton wool impregnated with androstenone. Thus, people could come up to the poster and test themselves. Another colleague hung three-dimensional viewers from his poster; these enabled visitors to appreciate the depth perception of his electron micrographs. In 1983 I attended a large international congress in Mexico. On the last day of the meeting, when most participants were burned out by the vast number of presentations, I noticed in a corner of the poster hall a crowd that had gathered around what should have been an empty board. When I approached I saw that a poster had been improvised in the space. It consisted of multicolored drawings of an *Apatosaurus* and a *Tyrannosaurus rex,* created by the 5-year-old son of one of the meeting's participants. Effectively titled "Macromorphology of Dinosaurs," it was the most popular poster of the congress!

4

Delivery

Voice Control and Eye Contact

No matter how spectacular the information or how beautiful the slides, **a scientific presentation, like any oratorical endeavor, stands and falls with the delivery. Effective use of the voice, eye contact, posture, gestures, and enthusiasm distinguish a routine presentation from a memorable one.** Some purists will claim that in scientific presentations it only matters what you say, not how you say it. Nothing is further from the truth. Delivery is important in establishing the impact a scientific presentation makes on the audience, and speaking skills can be determining factors in scientific careers. The force of your delivery may make the difference between whether or not you get a job offer, whether or not your grant proposal is approved.

Delivery, not content, often makes the lasting impression. General Charles de Gaulle practiced his speeches in front of a mirror to underscore his eloquence with carefully orchestrated gestures. When dropping in popularity, Jimmy Carter practiced clenching his fist to show determination and restore his leader image. Strength as a speaker is not only a talent; it can be an acquired skill. Most recent presidents of the United States speak in very similar speech patterns—a somewhat patriarchal, soft-spoken yet determined voice with a slightly majestic undertone. A Presidential manner of speaking has evolved that successful candidates need to adopt to appeal to the expectations of the electorate. This manner of speaking is a characteristic attribute of the office—not an inherited trait of those destined to occupy it.

The characteristics of delivery in terms of voice control can be separated into several interrelated properties: sound, volume, speed, and rhythm. The last term refers to such issues as monotony and emphasis. Of these four vocal characteristics, sound is the least problematic, since audiences will rapidly get accustomed to almost any sound. The only time sound can be a severe problem is in the case of certain accents. My Dutch accent has stayed with me as an inseparable companion, even though I have lived in the United States for 16 years. The moment I greet a stranger with a simple "Hi," the first question I am asked is invariably "Where are you from?"

It is virtually impossible for an adult to get rid of an accent. In most cases, accents present no difficulty; they may even be an asset, attracting the attention of the audience. Some accents give an air of sophistication. It would be unusual to hear the conductor of a major symphony orchestra speak with an ordinary Ohio twang. For the maestro, his European accent is a valuable aspect of his image. Sometimes, as in the case of Presidential speech patterns, accents can be acquired as a distinguishing feature. Some Oxford scholars have learned to speak English as if they are holding hot potatoes in their mouths. They were not born that way!

I have taught many foreign students the art of oral scientific presentation. Whereas most European and South American students had few problems, Chinese, Korean, and Japanese students often suffered from their accents, which can make them hard to understand and actually interfere with their attempts to communicate with the audience. They find it difficult to adjust to the many sounds and half-sounds of the English language and to its structure, which, unlike Japanese and Chinese, makes a clear distinction between singular and plural. The confusing use of singular and plural forms and frequent inappropriate deletions of "the" or "a" in front of nouns, combined with a heavy accent in which the "r" and "l" have merged as one sound, often tax an audience's patience.

When considerable amounts of the limited store of listener energy must be invested in deciphering the words themselves, less energy is available to perceive the concepts that the speaker is trying to communicate.

Although accents will never disappear, they can be modulated to within manageable limits. I remember one Korean student who had a difficult time not only with his accent, but with the English language as a whole. He persevered, and several years later this initially unintelligible student gave a polished and eloquent thesis defense. To improve their accents, foreign students should mingle in everyday life with native English speakers. This encourages them to communicate in English and, ultimately, will enable them to *think* in English. Many foreign students have learned to speak English by translating word-for-word from their native language. This gives rise to an unnatural, stilted speech pattern; moreover, expressions translated from another language into English do not always accurately reflect their intended meanings. Once a foreign speaker becomes thoroughly immersed in the English language through everyday use, he or she stops translating and starts thinking in English. This always leads to a big improvement in accent and communication skills. A cassette or tape recorder can also be used to listen to one's own voice, paying particular attention to especially prominent features of the accent. Speaking the same text with gradual improvements into the tape over and over again will ultimately lead to the desired result: modulation of the accent into a perfectly understandable vocal sound.

Foreign speakers who have severe language problems but nonetheless have to give scientific presentations in English should do two things: (1) Rehearse and practice the presentation often, preferably with a friend who is a native English speaker, and almost learn it by heart. Although memorizing a lecture is often not desirable—it may interfere with the spontaneity of a presentation—being able to form understandable sentences overrides this concern in cases where the speaker has a serious language problem. **(2) Structure your slides in such a way that the images are able to convey most of the story by themselves, even if you are hard to understand.** This may lead to the inclusion of more text on slides than would normally be desirable, but again we have to compromise and shift the weight from the speaker's oral presentation onto the visual aids in order to achieve our minimum goal: to communicate an understandable story to the audience. My experience is that, given enough time and enough investment of effort, foreign speakers rival their native English-speaking colleagues in eloquence. After two years of training, several of my Chinese students were among the top performers in my graduate seminar.

Volume and speed are two characteristics of speech that can, if necessary, easily be modified simply by paying attention to them. I have dealt with many students who spoke so softly that only listeners in the front row were able to hear them. It is important to raise the volume of your voice when giving a seminar; I do not remember ever having listened to a scientist who spoke too loud. My advice to any speaker is to **use a microphone whenever one is available.** It makes life easier both for you and for your audience.

Speaking softly is often an expression of shyness. I had one soft-spoken student rehearse her upcoming presentation with me by placing her at one end of a long corridor and myself at the other end. She literally had to shout to make herself audible to me. The next day, when she gave her talk, she spoke up and projected her voice. The louder volume, although it still felt unnatural to her, did not bother her as much after having had to shout the previous day. Just getting her to speak louder turned her into an excellent speaker. A similar remedy worked well for another student with the same problem. The student seminars at Duke are usually held in a room that holds about 40 people. To rehearse his presentation, I took him next door to a large lecture hall with a capacity of several hundred and placed him on the enormous podium. I sat up in the back of the auditorium as far away from him as possible. Forced to speak louder and project his voice, he grew accustomed to the vast space in which he found himself. The next day, the intimate lecture room looked much more sympathetic to him, and like his fellow student, he spoke up and gave an excellent presentation. In other cases I have made a little cardboard sign that said "Speak louder" and flashed it at the speaker when necessary. This small seminar room, like many others, has a low ceiling, a loud ventilation system, and a noisy slide projector. The acoustics are terrible—often it is difficult to understand a speaker from less than 20 feet away. Projecting the voice and raising it to an almost unnatural volume are essential in order to be heard by those sitting in the back row.

Articulation and eye contact are the two most important components of voice projection. Take the time to articulate every word of each sentence clearly, while maintaining eye contact with your audience. Mumbling, whispering, and letting sentences fade should be avoided, unless deliberately planned. A sudden lowering of the voice for one or two sentences, while maintaining eye contact with the audience and articulating each word carefully, can be used as a device to attract attention. The audience feels that it is being let in on a secret and allowed access to intimate information. The lowering of the voice forces the audience to listen more closely. Needless to say, this device is effective only when used occasionally.

Fading of the voice at the end of sentences is a fault frequently encountered even among experienced speakers. It represents more than just an aesthetic flaw in delivery. **As with written text, the end of a sentence designates the "stress position." It is here that the audience expects to be provided with the most important information.** Note the different emphases among the following sentences:

1. In mathematics Michael received an A.

2. Michael received an A in mathematics.

3. In mathematics an A was given to Michael.

The first sentence stresses the importance of the grade that Michael received in mathematics (he received an A, unlike the lousy grades he got in his other courses). The second sentence emphasizes the field in which Michael received an A (it's not surprising for Michael to get As in other fields, but in mathematics? his least favorite subject?). The third sentence puts emphasis on the person who received the grade (Carla and Jamal and Sarah always get As in mathematics, but Michael?).

Consider another example of the importance of the stress position in both spoken and written language, using the following three sentences:

1. Albert Einstein discovered the relationship $E = mc^2$ as the basis of his relativity theory.

2. The basis of Albert Einstein's relativity theory is the discovery of the relationship $E = mc^2$.

3. The relationship $E = mc^2$ was discovered as the basis of his relativity theory by Albert Einstein.

Again we can see the effect of the stress position. The first sentence draws our attention to the relativity theory, whereas the second sentence emphasizes the relationship $E = mc^2$ and the third sentence focuses our attention on the fact that Albert Einstein deserves credit for the discovery of his relativity theory. Both in written and in spoken language, important information should be placed in the stress position, where the recipient expects to find it. In written language, the stress position can be further emphasized by punctuation or changes of font. In spoken language, information contained in the stress position can be accentuated by alterations in the voice, such as an increase in volume or a pause, and underscored by gestures.

When a speaker's voice fades at the end of a sentence, the audience receives a contradictory signal. They expected the important information to arrive at the end of the statement, but instead that information is deemphasized by a lowering of the voice, leaving the listeners' expectation unfulfilled. When the last words of sentences are repeatedly inaudible or delivered with a fading voice, the audience becomes frustrated, like children who may look at the ice cream but are not allowed to have any. Consider delivery of the first sentence from our

Einstein example above by two hypothetical speakers:

1. Albert Einstein discovered the relationship $E=mc^2$ as the basis of his *relativity theory*. [increased volume at the italics]

2. Albert Einstein discovered the relationship $E=mc^2$ as the basis of his *relativity theory*. [fading of the voice at the italics]

In the first case the expectation of the audience is fulfilled loud and clear, whereas in the second case many members of the audience may never learn of what $E = mc^2$ formed the basis. The speaker promised to deliver information, but never fulfilled the listeners' expectation. Speakers who constantly allow their voices to fade at the end of sentences lose the audience's interest. Fortunately this detrimental habit can be corrected easily once the speaker has been alerted to the problem if he or she pays special attention to articulating the last words of every sentence until the habit is broken.

Mumbling and letting sentences fade can also be related to the speed with which one speaks. **Nervous, hurried speech often leads to inaccurate articulation. Take your time and**

do not speak faster than your normal conversational speed. When we take our car into the garage to be fixed, we explain very carefully to the mechanic the problems we have experienced with our miserable vehicle and what we would like to have done to it. The same careful speech pattern is appropriate for scientific presentations. One of my students used to speak with dazzling speed, so fast that it was remarkable that the words had time to take shape before they left her mouth. She also left no breaks between her sentences. I worked carefully with her, using audiotapes and rehearsals to slow down her speech. Again, I made a little cardboard sign that said "Speak slower" and held it up to her whenever she showed the inclination to speed up. She was initially tortured by the urge to exceed the imposed velocity, which to her felt impossibly slow and unnatural. The moment she slowed down, however, she became an excellent speaker. Moreover, she found that she had no problem covering the same amount of material. The speed of her thoughts was allowed to catch up with the speed of her speech, and as a result she expressed herself more effectively. She has permanently modified her style of speaking and remains a superb lecturer.

Despite good volume, clear articulation, and a comfortable speaking speed, a speaker can still deliver a soporific story if the rhythm of the voice is not controlled. **Monotony is the greatest enemy of a scientific presentation.** I always advise students, unless they have a severe language problem, never to learn their seminars by heart or read them verbatim from notes. This generates the "tape recorder syndrome," where a speaker rattles off a prepackaged script. Once a speaker recited a prelearned seminar without noticing that his message had gotten "out of sync" with his slides. The effect, unbeknownst to him, was hilarious. On another occasion one of my students learned his presentation by heart. He was a foreign student with a language problem, and I had recommended that he try to memorize at least parts of his lecture. Everything went fine until suddenly someone in the audience asked a question in the middle of his presentation. Not only was he unable to answer the question, he could not continue his talk. The tape had stopped and he had gone blank. He had to rewind and restart at an earlier point, repeating part of his prelearned message verbatim to be able to pick up where he left off. Although this is an extreme example, reading a written seminar or reciting a prelearned presentation hinders spontaneity.

Breaking the monotony is most effectively accomplished by placing emphasis on important phrases throughout the presentation. There are three ways in which emphasis can be placed in spoken language: (1) changing the volume of the voice; (2) repeating words or phrases; and (3) pausing. Combining all three devices is often extremely effective. For example, to motivate the audience to pay attention when I lecture on olfaction, I start the presentation by stating that "olfaction is the molecular recognition of chemical signals, which carry information about the localization of food and the availability of reproductive partners, and which are crucial for the survival of most animals." To emphasize this point and arouse the audience's interest, I change volume, repeat phrases, and pause, as follows: "Olfaction . . . olfaction is the **molecular recognition** . . . the molecular recognition of chemical signals . . . chemical signals which carry information . . . chemical signals which are **crucial** for the survival of most animals. . . ." (bold type indicates increased volume). Repetition like this should not be used excessively throughout the presentation, but can be a powerful tool to create a sense of excitement, urgency, and importance at certain key junctions in the talk, especially near the beginning or the end.

Pauses are a highly underrated, powerful means for impressing a point on the audience. A moment of silence can achieve more than a thousand words. We are by nature uncomfortable with silence. One of my former mentors, a psychiatrist by training, was a master in the use of silence. When you entered his office, he would offer you a seat, sit down across from you, and simply look at you in silence. This created an uncomfortable atmosphere that forced his visitor to say something, anything! Soon, without realizing it, you would be telling him all your plans, worries, and secrets. Similarly, during a scientific presentation a gap in the steady flow of words creates a slight discomfort, an air of expectation, which causes the audience to pay attention. Most importantly, the pause allows the last words to sink in. It gives the audience time to fully appreciate the information that it received just prior to the pause.

One of the most effective and most underused techniques for emphasizing key statements during a presentation is the double pause. The statement that the speaker wishes to emphasize is both preceded and followed by a pause, isolating it from the continuum of spoken text. The first pause serves to arouse the audience's attention in preparation for the upcoming statement, whereas the second pause allows the message conveyed by this statement to sink in. Consider, for example, a sociologist who has been invited by a local government agency to give an informative briefing on alcoholism in the community. A section of the sociologist's presentation near the end of the talk might run as follows:

The data presented on the last slide clearly indicate that government spending to combat alcohol addiction in this state has consistently lagged behind the per capita expenditures allocated to deal with this problem in other states. Furthermore, there is a severe communication problem between local government agencies and volunteer community organizations. Alcohol addiction is a major social problem in our community. Each year more than one hundred alcohol-related fatalities occur on the roads in this city alone. Drunk driver convictions have skyrocketed. However, alcohol addiction is not a crime, it is a disease. The government has to work together with community groups to cure this disease. This, of course, requires the availability of additional financial resources.

Let us now suppose that the sociologist wants to emphasize particularly the drunk driving problem, to stress the message that lack of adequate government funding is in part to blame for alcohol-related fatalities. Her presentation would be identical except for the insertion of a double pause, as follows:

The data presented on the last slide indicate that government spending to combat alcoholism in this state has consistently lagged behind the per capita expenditures allocated to deal with this problem in other states. Furthermore, there is a severe communication problem between local government agencies and volunteer community organizations. Alcohol addiction is a major social problem in our community. [pause] Each year more than one hundred alcohol-related fatalities occur on the roads in this city alone. [pause] Drunk driver convictions have skyrocketed. However, alcohol addiction is not a crime, it is a disease. The government has to work together with community groups to cure this disease. This, of course, requires the availability of additional financial resources.

The perception of a state official in the audience is as follows (pauses are again indicated by square brackets): "words, words, words, words, words, words, words, words, words, words, words [Hey, what is happening? The words have stopped. Why, what is going on? (Attention has been roused)] each year more than one hundred alcohol-related fatalities occur on the roads in this city alone [More than one hundred. That is about two fatalities every week. That's a lot! That's truly shocking] words, words, words, words, words." If the sociologist instead wanted to emphasize that alcoholics should be considered patients rather than criminals, a double pause could be used to isolate the statement:

[pause] However, alcohol addiction is not a crime, it is a disease. [pause]

Now the state official in the audience will leave the room with the idea that alcoholics should not be punished but rather treated as patients suffering from an illness. The verbal text of the presentation is the same, but the simple use of an emphatic double pause has changed the listener's perception dramatically. Notice also that the use of the emphatic double pause has allowed the sociologist, who was invited merely to present an informative briefing, to express a political opinion very effectively ("government negligence is in part to blame for alcohol-related fatalities" or "more drunk driving convictions will not solve the deeper underlying problem of alcoholism"). The presentation may have contained extensive documentation on alcoholism among different age groups and economic classes, data on alcoholism in minority groups, and an evaluation of the regional prevalence   of this problem. Yet the few words flanked by the double pause will receive a disproportionate amount of attention, although they represent only a small percentage of the presentation. The double pause is a powerful tool because of its subtle simplicity. It can also be used easily by shy speakers who feel inhibited to raise their voices or use gestures for emphasis. A slight increase in volume and a deliberate slowing of speech can further enhance the dramatic effect of the double pause.

Like the audience, speakers usually feel uncomfortable with silence. When gathering their thoughts between sentences, they often say "um." This is a very distracting habit and should be unlearned, if possible. **Plain silence is preferable to mere noise.** Many speakers also insert meaningless words like "basically" ("What this slide *basically* shows—") or "specifically" ("We can *specifically* conclude that these effects are *basically* artifactual"). I listened to one seminar speaker who not only talked with great speed, but after every other sentence introduced the word "okay?". He obviously did not expect an actual response from anyone in the audience, but employed this little phrase as a filler between sentences. He used it so often that it became ridiculous and made people laugh. If he could slow down and get used to incorporating moments of silence between his sentences, the urge to say "okay?" every minute or two might fade away. Moreover, brief pauses between sentences allow you time to think and clearly formulate the next statement. **Slowing down is a remedy for 90 percent of most speakers' problems.**

Finally, as mentioned earlier, eye contact is one of the components of a strong delivery. Many speakers avoid eye contact with the audience. Yet **eye contact and posture contribute crucially to charisma and presence.** We communicate with our eyes as much as with our words. A speaker who exclusively addresses the projection screen or has his eyes permanently cast down to the lectern or the floor fails to link up with the audience. **Looking straight at members of the audience establishes the notion that you are talking *to* them, not just *in front of* them.** Eye contact should not fix on one individual but should be supple, move around the audience, and involve all of the listeners.

171

Often it happens in our department that throughout their presentations speakers fix their eyes on the chairperson and make the rest of us in the audience feel as if we are intruders in a private conversation.

Eye contact not only helps voice projection and communication, it also generates confidence. He who can look his fellows in the eye is instinctively perceived as having a clear conscience and is considered trustworthy. Information offered with eye contact subconsciously gains more credibility than the same data delivered by a speaker who avoids looking at his audience. So, let us look the jury straight in the eyes and say in a clear, emphatic voice: "Here I am! I will tell you the truth, the whole truth and nothing but the truth!"

Posture and Gestures

When we are on the podium in the spotlight, we become performers. Actors are acutely aware not only of voice, but also of posture and gestures. Posture and gestures, like eye contact and vocal inflections, contribute to the stage presence of a speaker. This in turn will help determine whether the presentation will linger in the audience's memory or quickly be forgotten. I rather enjoy being "on stage"—you may as well, since there is no alternative. On stage I let my inhibitions go and speak freely, deliberately using my voice, body position, and gestures to give a *performance*, not just a lecture.

To make a lasting impression you have to be visible. Speakers tend to be glued in place behind the lectern, especially at scientific conventions where each speaker has only 15 minutes to give a slide presentation (Figure 21). The audience can listen to just a few presentations before its attention starts drifting. When scheduled in the middle or near the end of such a session, it is important to *perform* rather than just talk, in order to be remembered as more than one of the many faces on the podium. **Stand straight up.** You, the speaker, are the leader, the general who gives the commands, and need to look the part! Step out and away from the lectern and avoid talking to the screen (Figure 22), the slide projector, the floor, or the pointer (Figure 23). When pointing to the screen, interrupt your eye contact with the audience only momentarily and reestablish it as soon as possible (Figure 24). **Do not be stationary—change position occasionally and move around the podium.**

(Text continues on page 178)

Figure 21 This speaker, hidden behind the lectern, lacks stage presence.

Figure 22 Talking to the board or to the screen, this speaker is detached from his audience.

Figure 23 Avoiding eye contact by talking to the floor, this speaker slouches.

Figure 24 A dynamic speaker presenting data.

This, again, helps keep the audience's attention and also allows eye contact with different sections of the room. However, **avoid distracting mannerisms like swinging the pointer aimlessly around.**

Many novice speakers, especially when they are apprehensive, do not know what to do with their hands. They often put one hand in a pocket, while fiddling with or leaning on the pointer with the other hand. I have seen speakers with shaky hands try in vain to focus the laser pointer on the screen, while the untameable red dot darts around amplifying their tremors. (Thereafter, when they forget to turn the pointer off, the red dot drifts up to the ceiling or down onto the podium, taking with it the audience's attention.) I have seen very insecure speakers give entire presentations with hands clasped behind their backs. Others hold onto the lectern. None of these are natural or relaxed positions. **Speaking with a hand in your pocket looks sloppy and unattractive.** Far from nonchalant, this posture reflects tension and insecurity. You can, however, make positive use of your hands when you want to emphasize phrases during your presentation (Figure 25). As pointed out before, **gestures**, such as the clenched fist to show determination, **can under-**

Figure 25   Stressing a point.

Figure 26 An emphatic pause. Notice the sustained eye contact and the momentarily frozen gesture, which underscore the speaker's last phrase.

score spoken language. Consider the *voila* gesture, the outstretched arm that extends an invitation to the audience to inspect the information on display; the raised hand and forefinger that stress a point; the hand motions that underscore words, palms up reaching out to the audience. Finally, immobility of the speaker with a frozen gesture and intense eye contact can punctuate an emphatic pause (Figure 26).

As Shakespeare said, "All the world's a stage, / And all the men and women merely players." When giving scientific presentations, we all are actors. The scientific community is our audience, our careers stand or fall on the applause, and the best performers get to play the leading roles.

Enthusiasm:
The Indispensable Ingredient

A few years ago, I attended an international convention in Amsterdam and listened to a particularly dull speaker. One would expect anyone to greet the opportunity to give a presentation to the international scientific community with excitement and enthusiasm. In this case, the reality was different. Seldom have I listened to a less vibrant or inspired lecturer. He could have stepped into any funeral parlour in the city and continued his presentation; nobody would have frowned, since most eulogies are more animated. I have had only one similar experience: at a meeting in Florida, a researcher managed to present very elegant and important experiments in the flattest and most boring fashion. Needless to say, both managed to convey a deep sense of depression, but ignited no appreciation for their scientific accomplishments.

Most speakers show at least some enthusiasm for their work. It is impossible to motivate anyone in the audience to pay attention if even the lecturer is not excited about the presentation. One speaker began his seminar by apologizing to the audience for the fact that he was missing his best slides, which one of his students had borrowed and forgotten to return. In addition, he confided, the beginning of his talk was bound to be boring since he would have to cover some very basic aspects of his field. He informed his listeners that if they wished to sleep through the first half of his seminar, they should certainly do so; he promised to wake them when he was ready to discuss the more interesting findings of his research. Thus forewarned, many members of the audience followed his advice and did not wake up till the end of the presentation.

Enthusiasm cannot be taught. It is the spark from within, the belief that no matter how slight your contribution is in the grand scheme of Mother Nature, it provides a glimpse into her secret treasure box—a small advance, but one that is there forever. Every scientist deserves to be respected in the conviction that the topic of his or her work is the most important issue in the universe. Nothing is a greater pleasure than to watch a truly enthusiastic speaker who intensely enjoys a moment in the spotlight for himself and his research.

Enthusiasm is often suppressed, especially by novice speakers, out of timidity or insecurity. Let these inhibitions go! General principles of how to become a better speaker can be laid down, as in this book, but in the end **genuine enthusiasm accounts for 90 percent of a speaker's success.** It is this spirit of excitement that inspires the next generation of scientists, that convinces the prospective buyer to try the product, that leads the management to give the proposal a shot. It is passionate absorption in our work, a commitment that cannot be expressed in monetary terms, that gives us the irrepressible energy to move forward in our research. True enthusiasm and excitement can light a spark in the minds of your audience and arouse their eager interest.

Answering Questions

"I want this talk to be informal. Please feel free to interrupt me at any time." I frequently hear naïve speakers extend this open invitation to the audience to disrupt the presentation at will. Not a good idea! Although a stimulating discussion with the audience should be encouraged and often is one of the most valuable aspects of a seminar, it is counterproductive to invite the audience to break up and disrupt the momentum of the presentation with *ad hoc* questions. **The speaker should attempt to control the crowd, permitting questions at *his* or *her* convenience.** You can either ask at an appropriate moment in the presentation whether there are any questions, or (preferably) take questions at the end of the talk. Compartmentalizing the event in this manner will allow the presentation to flow smoothly without disturbing your train of thought. Interruptions may require lengthy answers and become full-blown discussions, derailing your talk forever. Most audiences are disciplined enough to wait until the end of the presentation. Occasionally, however, someone in the audience will blatantly interrupt with a question. The speaker has two options: to answer it, preferably briefly, or to ask that the member of the audience please let the seminar proceed—you will get back to the question at the end of the presentation. If you opt to answer at that moment, you risk encouraging further interruptions. Try to delay the answer until after the conclusion of your talk, by which time the need for the question has often dissipated. **By making it politely clear that the audience should not interrupt, the speaker will discourage impulsive *ad hoc* questions and can focus on the presentation. In doing so, you will also establish control and authority.**

The discussion following the lecture is an extremely valuable opportunity for the speaker to show depth of knowledge and to engage in stimulating debate with the audience. Preserve the same poise in terms of clearly articulated speech, posture, and gestures during this period that you mastered during the presentation itself, rather than collapse into your off-stage persona. The performance is not yet over. **Always answer questions briefly and to the point.** Some speakers use a simple question as an excuse to lead into a continuation of their seminars. This obnoxious habit is annoying to the audience and adversely affects the perception of the preceding talk. Precise, short, and well-articulated answers (rather than protracted expositions) allow the discussion to retain the vibrant momentum established during the presentation itself.

Who is given the opportunity to ask a question? The selection is sometimes made by the host, sometimes by the speaker. It happens often that one or two members of the audience dominate the discussion. It is important to involve as many people as possible, however, and to provide the opportunity to ask questions to different listeners. **It is in many cases advantageous to repeat the question before answering it.** Not only does this give you extra time to reflect, but it also keeps your entire audience attentive without allowing the post-seminar discussion to deteriorate into a private dialogue between you and one or two vocal members of the audience. In addition, repeating the question gives you the chance to *rephrase* the question. This can be of strategic value when the question you face is hostile. Following a presentation on neural development in *Drosophila*, for example, a speaker was asked whether he really believed that studies of fruit flies had any direct relevance to modern medicine and human disease. The lecturer not only repeated the question but also reshaped and generalized it, informing the audience that the issue was whether studying organisms other than humans can advance our knowledge of biological systems and, therefore, ultimately benefit medicine. Rephrasing the question immediately made the audience realize that the criticism in the question was not profound. They understood that the question was unsophisticated and ill-conceived, since everyone recognized that in most cases it is difficult, impossible, or unethical to conduct controlled experiments on humans; model systems in other species are therefore essential. By repeating—and recasting—the question, the speaker turned the tables on his hostile interrogator even before he proceeded to answer.

There are two types of questions: those that are anticipated, which can readily be answered, and those that are unanticipated, which in some instances pose problems. Anticipated questions usually recur on different seminar occasions, and the speaker has standard answers ready for them. **It is always a good idea to be polite and gracious.** Tell your listener, even if the inquiry is obvious, perhaps obtuse, "That is an excellent question" or "You bring up a fundamental point" before answering the question. It makes you look good and makes the questioner feel brilliant. Everyone is happy. Sometimes an aggressive competitor with dissenting ideas may be in the audience. It is important to remain calm and composed when challenged by an opponent. Never be abrasive. It is often preferable to avoid a full-blown confrontation in front of the entire audience. Remember, the speaker is one of them—they are on the same side! In such cases, it is best to say: "I can certainly see your point, even though I disagree with it. Since this is a very complicated issue, which may lead to a long debate, it is perhaps better if you and I get together afterward to discuss it."

Sometimes an unanticipated question can be more than awkward. When I studied toward my doctorate at the University of California at San Diego, I had to defend a research proposal outside my actual thesis project as one of the requirements of the program. The committee consisted of three professors. One, a Nobel laureate, stared out the window during my entire, very carefully prepared presentation. Occasionally his head nodded and he seemed to doze off. I suspected that he found my efforts very dull and that I had put him to sleep. The other two committee members started firing questions. No problem! Then, suddenly, when I thought it was all over, the silent professor spoke up: could he ask one small question? It was a killer. He had put his finger on a basic flaw in my hypothesis. We are not often confronted with such unanticipated, brilliant questions, but it has happened to me more than once and I have learned the hard way how to deal with these situations.

Difficult, unanticipated questions require a moment of thought. One way for the speaker to buy extra time before answering is to ask the questioner to, please, repeat or clarify the inquiry. If a question cannot be easily answered, it is always a mistake to feign knowledge or to talk around the issue. I have seen many speakers dig themselves deeper and deeper into a hole because they were not prepared to say simply: "I am sorry, I don't know." This, or a statement like "This is a very interesting point. I have not looked at it this way before, and I would have to think about it for a while. Why don't we discuss it together afterward" are usually the most face-saving. Unlike the reply to an aggressive adversary exemplified earlier, this proposal should be sincere. It is of real importance for the speaker actually to follow up and seek a solution to the problem by discussing it with the person who asked the question. This way the speaker can be prepared for the same question, should it come up again during a future seminar.

Only one thing is worse than being asked a brilliant, unanticipated question: not being asked any questions at all. We may like to flatter ourselves that everything was so clear that nobody could think of a thing to ask, but this is unlikely. It is far more likely that we lost the audience. Our graduate students are privileged to receive honest, constructive feedback after each of their presentations, but frank criticism is hard to come by in the real world. The nature of the questions or the lack thereof gives us some information about the perceived quality of our presentations.

Absence of compliments can also be disheartening. On the other hand, praise may not always mean much, and in the end, we have to be our own judges. One colleague of mine will eagerly run up to any speaker and tell him: "You were magnificent! This was the best talk I ever heard!" Afterward most people will say that they "enjoyed your seminar," whether this is true or not. Many years ago I gave, by my own admission, a mediocre lecture on benzodiazepine receptors. A famous scientist who is a member of the National Academy came to my seminar and fell sound asleep in the front row. He did not wake until the modest applause disturbed his dreams. That evening he was among the faculty members who took me out for dinner. As he sat down across from me and unfolded his napkin, he said: "You were magnificent! This was the best talk I ever heard!"

Finally, let it be of comfort to the reader to know that the perfect speaker is rare indeed. I recall only a handful of seminars that defied any form of improvement. All we can aspire to do is learn from our errors and get better with every experience. The guidelines discussed in this book will help the reader to explore and master the art of oral scientific presentation. But eventually one must move beyond the rules. **The most important advice to remember is, *communicate* with your audience and convey *enthusiasm* about your work.** Then, if you are armed with a coherent story based on solid data, you can enter the scientific arena, flourish your words of wisdom, and dazzle 'em with style!

Recommended Reading

Booth, V. 1985. *Communicating in science: Writing and speaking.* New York: Cambridge University Press.

Cairns, J. 1989. Speaking at length. *BioScience* 39, 632-633.

Cleveland, W. S. 1985. *The elements of graphing data.* Monterey, CA: Wadsworth.

Cook, E. 1976. Oral presentation of a scientific paper. In *Scientific writing for graduate students: A manual of the teaching of scientific writing*, ed. F. P. Woodford, pp. 150–166. Bethesda, MD: Council of Biology Editors.

Council of Biology Editors 1988. *Illustrating science: Standards for publication.* Bethesda, MD.

Floyd, S. 1991. *The IBM multimedia handbook.* New York: Brady Publishing.

Gard, G. G. 1986. *The art of confident public speaking.* Englewood Cliffs, N.J.: Prentice Hall.

Gatto, R. P. 1990. *A practical guide to effective presentation: It's not just what you say, it's how you say it that gets results.* Pittsburgh: GTA Press.

Kenny, P. 1982. *A handbook of public speaking for scientists and engineers.* Bristol, U.K.: Adam Hilger, Ltd.

MacMillan, V. E. 1988. *Writing papers in the biological sciences.* New York: Bedford-St. Martin's.

Picket, S. T. A., Hall, B. E., and Pace, M. L. 1991. Strategy and checklist for effective scientific talks. *Bull. Ecol. Soc. Am.* 72 (1), 8–12.

Schlof, L. 1992. *Smart speaking: Sixty-Second Strate-*

gies for More Than 100 Speaking Problems and Fears. New York: Plume.

Schwier, R. A., and Misanchunk, E. R. 1993. *Interactive multimedia instruction*. Englewood Cliffs, N.J.: Educational Technology Publishers.

Sindermann, C. J. 1982. *Winning the games scientists play*. New York: Plenum Press.

Sindermann, C. J. 1987. *Survival strategies for new scientists*. New York: Plenum Press.

Staheli, L. T. 1986. *Speaking and writing for the physician*. New York: Raven Press.

Index